T0179928

# CATERPILLAR
# ASSOCIATION
## OF THE UNITED STATES

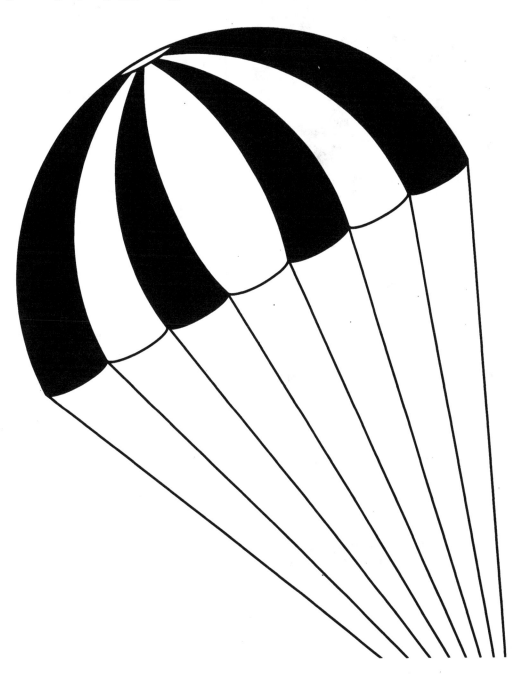

TURNER PUBLISHING COMPANY

*Lawrence Justin was hit by flak on June 22, 1944 over Cherbourg and parachuted into the English Channel. He was picked up by British Air-Sea Rescue and returned to the base unharmed.*

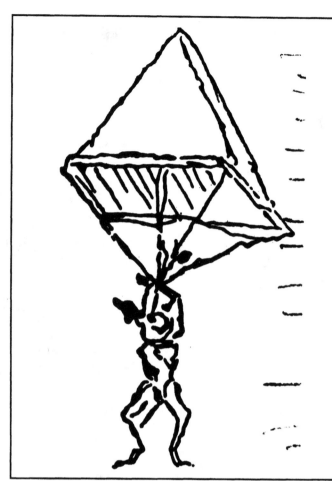

*Leonardo de Vinci's design for a parachute that he sketched in his notebook.*

## TURNER PUBLISHING COMPANY

Copyright © 1993 Turner Publishing Company

This book or any part thereof may not be reproduced without the written consent of the Publisher.

This book was compiled and produced using available information; the publisher regrets it cannot assume liability for errors or ommissions.

Libarary of Congress
Catalog Card No. 91-75220
**ISBN:** 978-1-56311-031-3

Limited Printing

Additional books may be purchased directly from the publisher.

*A soldier bails out and floats forlornly to the ground.*

# TABLE OF CONTENTS

*After landing, a soldier hauls in his parachute.*

*Johnny Brown*
*Fifty years after his bail out...*

This is not the history of the parachute, but rather the stories behind the men and women who were forced to use them in an emergency situation. However, the original invention of the parachute goes back to the year (1495) and is commonly attributed to Leonardo de Vinci, who illustrated it in his Codex Atlanticus, and said, "If a man were to have a tent roof of linen 12 breccia broad and 12 breccia high, he will be able to let himself fall from any height without too much danger to him, as an object offers as much resistance to the air, as the air does to the object."

At the start of World War I our heavier than air pilots had no parachutes, but about 75 of our "Balloonatics" were issued one, 50 of them made one or more jumps. Toward the end of the war, German aviators were issued chutes. The late General "Billy" Mitchell finally succeeded in getting about 100 for our pilots—but too late as the Armistice was signed two weeks after.

The Caterpillar Association, actually was born in 1982, a group of self-saved men and women, whose words and deeds are recorded in this book. Any person who saves his life jumping from a disabled aircraft, with a parachute, can become a member.

Since our founding, the Caterpillar Association has functioned solely as a name. During World War II, and after my jump, I wondered about the people who made my chute, and some day I wanted to meet with them face to face, and thank them for saving my life. I am sure that many others feel the same way. In some way I believe that the parachute manufacturers who had provided the parachute to save our lives, deserve a lot of recognition.

Johnny Brown
Caterpillar Association

Lt. Catlin's crew, 743rd Sqdn., 455th BG. Topeka, KS, May 1944. L to R, standing: Sgt. Lycan, Lt. Paurice, Lt. Greenquist, Lt. Catlin, Lt. Singer and Sgt. Linneweh. Kneeling: Sgt. Mattson, Sgt. Bodenhorn, Sgt. Skinner, and Sgt. Mazoros.

# SPECIAL STORIES

## MISSION TO MUNICH
*by Bob Greenquist, Navigator*
*on Leakin Deacon - June 9, 1944*

**I** was the navigator on the Leakin Deacon, 743rd B.S., 455th B.G., 15th A.F. on June 9, 1944 on a mission to Munich. I knew we had flown over the Alps in Austria and Italy and thought we had also flown over Switzerland. However, after putting together our collective remembrances and examining a few maps, I doubt we flew over Switzerland.

We headed for Switzerland and entered Austria to the west of Garmisch-Partenkirchen. I believe we went into the Fern Pass (3967 ft.) and flew SSW. At our airspeed and with out inability to gain altitude, decisions had to be made so fast, as we flew blind through mountain passes, that maps were useless. Al and Buddy watched the wingtips while I stood between them visually picking the route ahead. There were times when both wingtips were 10 feet or less from the mountains and there were times we ran into deadends. We gambled - and won - but we were very, very lucky.

My guess is that we flew through the Resia Pass (4936 ft.) near the intersection of the Swiss, Austrian and Italian borders. Just south of the pass is Lake Resia, whose outlet is the Adige River. The Adige flows south, then east to Merano and south again to Verona before turning east and emptying into the Adriatic. I believe we followed higher valleys roughly parallel to the Adige Valley toward Merano but that we turned south before reaching Merano to avoid the larger cities in the Adige Valley between Merano and Verona.

I don't remember our taking a vote to head for Corsica nor do I remember deciding to go there until we headed south along the valley which took us over Lake Garda. That decision was made on the flight deck. Over Lake Garda we dumped overboard some of our equipment. Shortly thereafter, as we reached the plain of the Po River, we bailed out. I don't know where I reached the ground, except that it was north of the Po River, nor do I recall the name of the town where I was picked up by Italians.

I know that I was taken to Verona by the Germans for extensive

*Six of the original crew members from the Leakin Deacon. Picture taken at Dayton Caterpillar Reunion. (Courtesy of Arthur W. Mattson)*

interrogation. I understand that Bob Skinner was taken to a hospital in Guastella and my guess is that we all may have landed within 10 miles of Guastella.

I was first out of the bomb bay and did not see any other chutes. Since Bob was first out of the back of the plane and appears to have landed south of the Po River near Guastella, my guess is that I may have been the only one to land north of the Po. The more I stare at maps, the more familiar the town of Casalmaggiore seems to sound and to look like the town where I may have been captured - but I really don't know. *Submitted by Arthur Mattson*

## ONE WAY TO BECOME A CATERPILLAR
### *by Randell S. Meyer*

This brief story takes place on 1 May 1984 over the Great Salt Lake desert northwest of Salt Lake City, Utah. I was part of an F-16 mission from the 421st Tactical Fighter Squadron, conducting training for one of our new fighter pilots. It was an interdiction mission, where two aircraft were to put bombs on a runway out in the desert. My two-ship was to act as "red air" and attempt to find and shoot down the strike aircraft before they got bombs on target. Due to some maintenance problems with my wingman's aircraft, the strike fighters took off on time while we were delayed on the ground. They fixed my wingman's bird, and we took off, flying the "speed of heat" trying to find and catch the other two fighters before they got to target.

We found the strikers about 30 miles short of the target on their way in and flew a low altitude, high speed intercept. As we began the intercept we were opposite heading of the strikers, at about 500 knots and 1,500 ft. My wingman was on my left side. We flew a conversion turn in afterburner to roll in behind the strikers and as we rolled out, my wingman came out on my right side.

As we chased the strike aircraft and prepared to shoot,

my wingman yelled over the radio in a tense voice "knock it off" (which in our language means stops all training, something's wrong). The tone of his voice concerned me, so I climbed (we were now down at about 300 ft., 550 knots), canceled afterburner, and crossed over my wingman to look at his jet. I saw nothing wrong, so I asked him what the problem was. His response: "I think you're on fire".

I looked at my gauges, but saw nothing unusual. I looked out over my shoulder and could see nothing but a wall of fire obscuring the tail of the jet. I continued my climb and reduced power some more, but it was a magnesium fire that wasn't about to quit. Soon I saw pieces coming off my jet, and I knew I had problems. The aircraft started a right roll at about 3,000 ft., which I couldn't correct; it seems the flight controls had burned through.

I figured there wasn't much heroism in staying with the jet so I called my position to a local radar facility and pulled the handle between my legs. Time really slowed for me as I remember to this day what every gauge read, and also remember waiting forever for the seat to give me the kick (actual time less than a second). Once I got the jolt, I saw the F-16 slowly falling away (actually I was rapidly going up the rail). I lost my sight as the wind blast hit me (I ejected at about 450 knots) but did not feel the wind blast, and the seat automatically separating from me.

The next thing I remember is waking up in the chute. I had a good chute with no problems. I accomplished a four-line jettison (which allows more steerability), and enjoyed the ride down. I hit pretty hard, and of course landed on a cactus, but was alive. I had a couple of substantial bruises and an injured neck but eventually got back into the cockpit. The jet did a spilt-s, flew into the ground at an 89 degree angle, and disintegrated. A USAF helicopter picked me up about 20 minutes later and brought me home to a warm welcome.

## A DAY TO REMEMBER
### *by John P. Mulvihill, Jr.*

My story is not one of noted excitement, but one that I will always remember. It started early in the morning of Feb. 19, 1945, from the base of the 828th B.S. (B-24's) to which I was assigned as a bombardier. The 828th was part of the 485th B.G. of the 15th A.F.

I had flown the previous day with another crew but the mission was aborted because of weather and I was surprised to be flying again, however this was my regular crew except for the navigator, a Lt. Feldman. The weather was fair but reports were good for the target area which was Graz. Other groups were going to Klangenfurt and Vienna. The flight in general was uneventful although we were hit by some flak that caused us to lag behind prior to reaching our objective.

The pilot, Lt. McKeon, finally decided to drop back and return to base after we lost an engine. I told him on our return we could make a run on Pula which was considered a secondary target and drop our bomb load, which we did. Immediately thereafter we lost our second engine, however we were headed toward Italy although losing altitude.

After a distance out over the Adriatic, the third engine started to act up. We then headed back towards the Yugoslaia coast with the hope we could make Zadar and bail out. However we were losing altitude quite rapidly, making it necessary for us to bail out then.

We were slightly off the coast and could see land and some islands below us. The engineer left first, followed by the navigator, myself, and radio operator. I noticed that Christenson had gotten out of his seat and was adjusting his chute to jump. It was about 2:05 p.m. at about 8700 ft. with the weather clear. I only counted seven chutes but I cannot say that they all did not get out as the plane disappeared behind some large hills. I did find out later that the navigator and myself were the only survivors.

When I jumped we were approaching some small off-shore islands. I did kick off my boots as I felt they might hinder me if I landed in water. I was fortunate to land on one of the small deserted islands about 35 miles northwest of Sibenik. Unfortunately I hit a large rock fence that stretched the length of the island, breaking both my ankles and severely cutting my foot. I was a little dazed but gradually came around and realized that I was hurt.

I cut a piece of my chute with my knife and used it as a tourniquet to stop the bleeding. This was about 2:40 p.m. in the afternoon. About 3:00 I heard a voice calling and saw Feldman, the navigator, coming towards me. He had landed at the water's edge and had seen me come down. He was not injured except for a slightly sprained ankle.

The island had some old shelters on it which we were to learn later had been used for sheep that had grazed there. He helped me get to one of them which offered some protection from the wind. As evening came it was very cold and we built a fire. We had seen some 38's fly over but they did not see us. So I spent a very cold and miserable night.

The next day was clear and a little warmer due to the sun. Feldman spread our chutes out with the hope that someone would see them from the air. We had also seen a Navy ship off shore but too far away from us. I had controlled the bleeding pretty well, although I was in a lot of pain and discomfort.

On the next island we saw some type of building or house. About 3:30 in the afternoon a B-17 spotted our chutes and flew over us very low to let us know they had seen us. About 15 minutes later we saw two men coming towards us from the other end of the island, they were saying "Deutchland" "Deutchland". Lt. Feldman's grandparents had come from Budapest and he had enough knowledge to communicate with them and make them understand.

They indicated that they had seen us land, but it was too late for them to come yesterday. They took us back to their island in a little boat where there were two women and another man. They caught some fish which they cooked and gave us some type of wine which really only made me thirsty. They claimed the water was not good to drink. They were very good and friendly to us and said they would take us to Sibenik tomorrow where I could get some treatment for my leg. I was skeptical as the boat was small and the water rough. I could not move around much and I was hurting badly and would not have had much chance if the boat got swamped. I did know how to swim if I could swim with my injuries.

At about 5:30 we noticed a P.T. boat coming (turned out to be a British Air Sea Rescue). They had been alerted by the B-17. The boat took us aboard after much difficulty as they could not get too close to the shore.

We bid our benefactors farewell, leaving them my knife, which I knew they wanted as it was a switch blade. The British took us to Vis and treated me the best they could with their facilities. After spending the night in the infirmary, a C-47 came and returned us to Bari and the 26th General Hospital where they set my ankles.

After about three weeks I was sent to the 17th General Hospital in Naples to await shipment to the States. I returned on the Mariposa, which is now the Momeric.

I was hospitalized in the States at Plattsburg, NY until about the middle of November, then was given a leave, reporting to Greensboro, NC on Dec. 6, 1946 to be reevaluated. Since the war was practically at an end, I was offered the opportunity to leave the service which I did.

# NIGHT MISSION TO KAISER-SLAUGHTEN, GERMANY
### by James Musat

On the night (about 1:30 a.m.) of 10 February 1945, we took off in an A-20, designated a F-3 for night photo missions, from Florence, Belgium; our mission to photograph bridge damage at Kaiser-Slaughten, Germany. At the target we met with intense anti-aircraft fire and experienced moderate damage. We had to turn away and shortly saw a long string of lights on the ground.

I decided to make this a "target of opportunity" and proceeded to drop seven photo flash bombs, again with intense anti-aircraft fire and search lights. It turned out to be the railroad marshalling yard at Alstadt, close to Saarbrucken, Germany.

Again we were chased off and decided to get back to base. Near Metz, France at about 5,000 ft. we developed severe icing and began to lose altitude fast. At about 2,000 ft. Bill hollered on the intercom, "Bail out fellas, bail out."

I immediately went out the two ft. by 3 ft. trap door in the floor and between the propellers. I tumbled two or three times and then pulled the ripcord (D-ring). The chute opened just about the time I hit ground. What saved my life, (besides the chute) is that it was raining and sleeting and I landed on an inclined mud bank along a railroad track.

I lay half conscious in snow then recovered and hid in a small railroad shack until daylight. I wrapped myself in my parachute for warmth. At daylight I heard a church bell and figured a town was that way.

After about an hour I was surrounded by "Thank God!" American infantry soldiers with M-1 rifles and taken to their headquarters. After a night and day of questioning they finally believed my story and were somehow able to make contact with my squadron, who sent a vehicle to pick me up.

Although both ankles were severely sprained (possibly fractured) and my lower back was very sore, due to the shortness of experienced combat crews, I was back in the night flying on the 12th of February and then almost every night until April 7th when the war effectively ended.

Although our gunner, Dick Holst, also bailed out when I did, Bill Wolfs was unable to release his canopy, apparently due to severe ice, and was able to fly back to our base at treetop level. Upon landing the canopy jarred loose and flew off. This saved the pictures I took, and they were so exceptional they were published in the March 1945 issue of "Impact" magazine.

## RADIO SCRIPT OF GENERAL ARNOLD'S BIRTHDAY
*Recorded by "War Report" 27.1.45*
*S.S. Men and Fortress Crew*

**C**ue Material: This is the story of the crew of "Little Joe Junior" - a Fortress which developed engine trouble when bombing Hanau. First one engine began to throw oil, which slowed the whole plan up and just after she reached the target area, flak hits put another engine out of action. The Fort managed to bomb its target successfully and then turned for home, badly slowed down and losing height; the bomb bay doors were jammed in the open position and one oxygen system was leaking, but the pilot determined to make for friendly territory.

This is the story of the co-pilot; 2nd Lt. Herber Drumheller of Pottstown, PA; the ball gunner, Staff Sergeant Nicholas J. Peters of Wyandotte, MI; and the tail gunner, Staff Sergeant Clarence W. Gieck of Long Beach, CA. All members of the 8th U.S. Air Force.

**Peters:** We were limping for home and for quite a time everything was quiet - there was no flak and no sign of Jerry fighters but suddenly our No. 2 engine started vibrating badly and the vibrations shook the whole ship. The pilot tried to throw the prop off by making a series of violent maneuvers; but then flak started picking us up and the pilot gave the order to prepare to bail out. In spite of all this our pilot remained perfectly cool and very alert; his evasive action was perfect. He'd fly along until he thought the enemy had got the range and fired, and then he'd turn to another course, and sure enough, the flak would burst just where we'd been. He did this over and over again; if it hadn't been for his skill we'd have been blown out of the sky. Finally they were putting up such a barrage that evasive action was too risky, so we got the order to bail out. The flak was very heavy now, and one of our other engines was reported by the waist gunner to be on fire. We were now pretty low - in fact about 11,000 ft. and we were being fired at by 20mm cannon as well as the heavy stuff. The first three men bailed out - then I jumped out the waist door and at the same time Gieck here, the tail gunner, went out the tail door. Gieck, you got down before me, you tell them about that drop.

**Gieck:** For the first two minutes we were shot at by small arms fire and 20mm. Sitting in our chutes and looking down we could see little red sparks and white puffs below us – it sounded like popcorn. We could see the three boys who'd jumped before us high above in the sky – still coming down, but they were going to come down too far behind the German lines. We were glad to see the rest of the crew had now bailed out of the plane.

The co-pilot hit the ground first and due to the jar he got he couldn't get up until I had landed, got out of my chute and gone over to him. He was still lying there dazed. Then Pete landed and when Herbie – the co-pilot – had recovered we three went into a huddle. We weren't sure where we had

come down, we only knew the general direction of our line I had a small compass so we set out in that direction, walkir west and hoping for the best. Suddenly from a thicket nearb we heard a voice call "mericanish", and that stopped us u very short! At first we thought we had run up against son French but when they called us towards them we saw the were Germans. There were two of them, one had a rifle an the other a pistol. Pete, you understand German, what di they say?

**Peters:** The younger German called out "Come on dow here." We hesitated, but finally seeing that they weren pointing their guns at us we went down into the thicket. Or of them told me he was wounded in the left leg, so I bandage it up for him. They hadn't had any water but they offered u bread and cigarettes. They told me that they had been in th thicket for three days, and we found out afterwards that the were two S.S. men. I asked one if he knew where our line were, and he pointed in the direction we'd been headin, "Are you sure?" I said, "Well that's where the bullet that h me came from," he replied.

I suggested we <u>all</u> try and make for the American line but they wanted to wait until after dark as they said th American boys over there were pretty trigger happy. The told us that the Americans knew they were in the thicket an when I heard that I said, "Look here, we're going to get goin now, if you want to you can come along with us." Gieck sai we ought to have something to show we're friendly, so th Germans produced a big white sheet and Gieck tied this on long stick, and so we all set off.

We made a pretty queer sort of party. We plodde through the snow, Indian file, I was first with the banner, the a German, then Herbie, then the second German and final Gieck here. It wasn't too pleasant – all the time we could he machine gun fire and shells were going over our heads in a directions. We spotted a town and made towards it hoping was in friendly hands. Just as we were on the outskirts we sa two Germans standing on the step of a house watching u near the house was a German command car. We took all th in and did some fast thinking and trying not to show ou desire to make a quick get-away, we skirted around the tow When we last saw the Germans they'd turned and gone int the house. Obviously they had seen the Germans with us sti carrying their weapons and not unnaturally thought we wer being brought into town under armed escort. Lady Luck wa with us! This was no place for us so we got out of town as fa as possible. Gieck took the lead now and he led us down th railroad track.

**Gieck:** We'd gone about three quarters of a mile dow the track when we saw a G.I. helmet pop out of a hole and very welcome American voice say "Drop the banner." W had walked in to one of our own patrols, and soon the guided us back to our own lines. Source: Radio script: BB London, England

## IT WAS THE INVASION OF FRANCE D-DAY
*by S/Sgt. Orvis Preston*

**I**t was the morning of June 6, 1944. I was the first one out c the sack on this morning, walking outside of our quonset-hu ooking at the sky. I stopped and went back inside, waking everyone and announcing that the invasion of France was on.

The first words I had been ask was if I had heard it on the radio. I said no, just take one look outside and tell me it's not so. With the sky literally black with planes, it was the most dramatic scene my eyes ever saw.

Being a crew member of a Douglas A-20, light bomber, along with the rest of the men in the quonset-hut, some 20 men. By mid-morning we had all been alerted for missions.

I was with the 410th B.G., 646th B.S. (L), flying out of Gosfield, Essex County, England on A-20G. Our crew was 2nd Lt. Russell L. Goodchild, S/Sgt. John Sanford, and my-self, S/Sgt. Orvis C. Preston.

Thirty-seven 410th Havocs took off at 2005 to bomb a marshalling yard located at Longpre Les Coeurs Sainte, near Abbeyville, France. Four aircraft failed to attack, with 33 dropping 168 500 pound general purpose bombs on the target at 2111 hours, with heavy, accurate flak reported. Capt. Robert W. Paulson, 646th B.S., led the formation. Bombing results show that the first box hit the target but results were unknown for the second box as no photo coverage was available. Thirty-two aircraft were damaged by flak. Goodchild, pilot, 646th B.S., was killed in action. His gunners, S/Sgt. John Sanford and me, bailed out and were captured being POWs.

Excitement of the invasion, my ninth mission, and to find out we had no generators working, the radio was out, but the plane was in the air and we were on our way to France. The ships in the English Channel were so thick it looked as if you could step from one to the other. Traffic was one way and that was over only anyone turning back for any reason would be shot down by our own ships or troops. After the bombing the planes were to fly around the invasion and come back up from the southern most England. Our training was air support for the ground troops, to this point it had been medium altitude bombing.

For me the trip was somewhat shorter. At 2113 hours our bomb bay was ablaze, with fuel in our tanks. With no radio or intercom the next best was a parachute. I was the first one in the air and had been reported as one parachute, with the possible second parachute.

Free fall was about 500 ft. into a tree that broke my fall. Jumping up and down pulled the chute from the tree dropping me into a slue, chest deep in water and cattails. Tramping the chute under my feet I started to climb out of the slue which the bank was with in my grasp. I had all my flight gear on, was wet from the water in the slue. I was trying to get out of the area, but within 30 feet I pulled up my ear phones and could hear voices. Thinking maybe it was French, I stepped behind a small tree, started looking around and could see heads all around me. I was in an open field except for the tree I was behind and the slue I had just left. After a short time I walked out with my hands up and hoping they were French.

Many guns were trained on me, it sure was hairy as they were only kids, looking like about seven-year-olds to me. They had the guns, now what happens was funny, with my flight gear on the frisk was on, everything was a booby trap to them, there was my flak jacket, flight jacket, escape kit and my 45. This was the Hitler Youth and they kept the troops in line.

I was marched about a mile and a half and then into some pine woods where a log cabin with SS troops had an office. The SS tried to interrogate me and go over everything I had.

A piece of shrapnel had pierced my jaw, with my face swollen it was hard for me to talk or say anything and if I did it was hard to understand what I was trying to say. There was a hole through my jaw and into my mouth, I could stick my tongue in the hole and could rub it with my finger.

Going through five transit-camps, I was sent to Staglas Luft IV where I was kept until February 6, 1945. February 6th we were marched out of Luft IV as the Allies were closing in on our camp. The roads and farmers' barns were our camps for the next 87 days.

Attacks by our own planes even strafed the marching columns, killing some of our men. At one point I was so weak while on the march, they left me on the side of the road. Two English lads with a German guard came along helping me. That night about 2300 hours I rejoined the group that had marched off and left me. It was May 2, 1945, when a jeep with an English lieutenant and a Scottish sergeant waved a rifle at about 1000 hours. I was FREE!

# A LIFE SAVED FOR THE CATERPILLAR ASSOCIATION
### by Kenneth Hawk-Slaker

On the 9th of September 1948, I reported TDY to the 60th Troop Carrier Group at Wiesbaden Air Base from my home base at Furstenfeldbruck near Munich, to participate in the Berlin Airlift. After several of we newly assigned pilots were briefed by flight operations on flight procedures in the air corridors to Berlin, the Intelligence officer made this closing statement: "The Russians say that they will shoot down any aircraft that strays out of the Berlin air corridor, and that captured airlift pilots will be treated as spies. This is a serious threat to our pilots and if you should find yourself downed in the Soviet zone of Eastern Germany, we cannot say that you should turn yourself in, or that you should try to escape. There is no firm policy on this and it would be up to you as to what action you would want to take."

My first four flights to Berlin were at night and in weather. The weather was really bad the evening of the 14th when Lt. Steber and I reported for two scheduled flights to Berlin. Steber was an experienced C-47 pilot with the 60th TCG and because of the weight factor, we two were the entire crew. We never flew the same C-47, and carried a variety of cargo such as flour, sacks of charcoal, and other foodstuffs. We received our aircraft number and sloshed along the flight line, looking for our aircraft. When we spotted our Gooney Bird, I had a sinking sensation in the stomach. The faded pink paint on her surface exposed it as a relic from the African campaign in WWII. Since it was due for the graveyard, we gave it a thorough inspection, kicked the tires and climbed aboard.

When I sighted the cargo of 55 gallon gasoline drums, I visualized this crate blowing up from the fumes. I eased my 6'3" frame into my backpack chute, while Steber lay his chute, still buckled, near the rear exit. Takeoff was normal and we bounced through the weather to Berlin. Both of our faces cracked into broad smiles when GCA brought us into sight of the runway. We were unloaded within a few minutes and our flight back to Wiesbaden was instruments all the way.

For the second flight to Berlin, a German crew loaded flour into the aircraft, while our own refueling serviced both the main and auxiliary tanks for six hours of flying time. Our morale was a lot higher for this second trip to Berlin, what with a less dangerous cargo and confidence in the pink C-47 to fly. We were cleared for takeoff, to climb and to maintain 6,000 ft. altitude at an indicated airspeed of 140 mph. The takeoff roll was longer than usual, and after becoming airborne, we had to increase the RPM to 2400 in order to maintain climbing airspeed. We leveled off at 6,000 ft. and had to use increased RPM and manifold pressure to maintain the required airspeed for lateral separation from the aircraft behind us at the same altitude. It was extremely heavy rain and water was dripping into the cockpit, wetting our flying clothes. We reported over the Fulda beacon, turned to the magnetic heading of the south corridor into Berlin, and settled down for a rough and wet flight to Berlin. Steber was to fly the first half of the flight and I the second half of a two hour flight plan. I closed my eyes and attempted to relax before my turn at the wheel.

Twenty minutes into East Germany and without any warning, both engines quit simultaneously! The sudden loss of forward momentum caused both of us to lurch forward into the wheel, and thinking that Steber had pulled the power off for some reason, I asked: "What's wrong?" "I don't know," he replied. I immediately cross checked the instrument panel. Fuel pressure showed in the green and outside temperature was 12 degrees above freezing. All four tanks showed adequate fuel. "Change to auxiliaries", Steber shouted. We had taken off and flown on main tanks until the engines quit. I operated the hand wobble pump to insure fuel pressure but the engines did not respond. The staccato of the heavy rain on the metal skin was very audible and we were dropping through the turbulent black sky like a lead sinker.

I rechecked master switch on, magneto switches on, and fuel tanks and pressure, but the engines remained silent. "It's just some damn little thing if we could only fine it," shouted Steber. "What is the highest terrain in this area, Slaker?" asked Steber. "4,200 ft.," I replied. The altimeter was now at 5,000 ft. and Steber shouted: "Go back and open the cargo door and lay out my parachute while I go through emergency procedures again."

I scurried over the bags of flour, my parachute harness getting caught several times on the tie down rods that extended above the sacks of flour. I attempted to open the cargo door by the inside handle, but it was jammed. I pulled the emergency release handle and the door was sucked into the black void with a swish. I unbuckled Steber's parachute, laid it on top of a sack of flour, and crawled back into the cockpit. Steber had just finished going through emergency procedures for an engine failure without success. Indicated altitude was now at 4,400 ft. and the rain was leaking into the cockpit like Niagara Falls. "We are going to have to bail at 4,000 ft. unless these engines start," shouted Steber.

In a final and desperate effort, we turned master switch off and on, mags off and on, changed tanks, but the propellers continued to windmill. At 4,000 ft., Steber cried: "Lets go," and we scampered over the sacks of flour to the rear exit. I

grabbed Steber's chute and held it for him to slip into. I shouted: "For God's sake, Slaker, jump... we are going crash any second."

I immediately turned and fell headfirst through the do opening. I pulled the ripcord as my feet left the floor of t fuselage and the chute deployed at once. As I swung u wards toward the disappearing C-47, I watched for Steber bail out, but before he could appear in the doorway, I start to descend back down through the arc and hit somethi very hard on my backside. A million lights flashed in n eyes, and my last thought was: "This is it." And then the was nothing.

It seemed quite normal for me to be lying on my back a potato field, the rain falling hard and cold on my face. T sky was a solid grey and all was silent in this wet worl Amnesia is a terrifying experience after you are no long suffering from it. The sudden shock of remembering c result in a variety of mental and physical reactions. How lor I laid there, I don't know, but without any preview, n memory suddenly returned and my body trembled from tl shock. I broke out in a cold sweat but I recalled reading abo post accident shock reactions, so I knew what was happe ing. When the tremors subsided, I tried to stand up, b discovered that I could not move from the waist down. Tl back of my head was bloody and I had cut my tongue with n teeth when I hit the ground.

Without a doubt, we were not more than two or thre hundred feet above ground when I bailed. I raised my hea and looked over the drab countryside, but could not sight ar crashed and burning aircraft. The feeling started to return my legs and the words of the Intelligence officer came back me: "What action you take is up to you." I had no desire to b the first airlift pilot to be tried as a spy. I decided to make a effort to sneak across the so-called iron curtain.

"You have 20 minutes to brief General Tunner on th highlights of your experience from takeoff at Wiesbaden t your escape across the border," were the words of his execu tive officer upon my arrival at Airlift headquarters. I spen one hour with the general, informing him of my experienc and answering his questions. He informed me that I was nc to talk to the press, that the PIO would release a story that had walked out, in order to protect the Eastern Germans tha had given me assistance in crossing the border.

## DROPPING WITH ONLY ARM IN RING

*by John Stevens*

I was a crew chief in the Second Troop Carrier Squadror veteran transport outfit, in China-Burma-India Theatre. I July 1945 I was heading in a C-46 towards the foothills of th Hump, some 65 miles southeast of Chabua, over the rugge first ridge. The aircraft's right engine sputtered at about 7,00 ft. Seconds later the radio operator tore past me, grabbed parachute and opened half the cargo door.

I made my way to the cockpit in order to offer my service to the pilot but perceived that there was nothing I could de The pilot was yelling, "Bail out, bail out!" I retraced my step toward the rear of the plane and pulled a parachute from it rack. I was unable to remain on my feet because the plane wa buffeted about so badly by the up and down drafts.

I then stretched out on the floor, full length, and attempted to wriggle into the chute. This also proved futile. In utter despair, I hooked my arm through one of the loops which enamate from the seat of the chute, and pondered vaguely, the next step in this grotesque nightmare.

In a minute I felt myself falling in space. I realized my only hope of surviving was hooked in the crook of my arm. I twisted and turned to grope for the ripcord release and found it. I yanked it and miraculously the chute slowly began to unravel. The slowness was a marvel for if the big nylon blanket had blossomed forth in one grand jerking operation (as it generally will) the tremendous pressure exerted would have torn my arm from my socket.

It was impossible to control my descent in any way and I watched helplessly as blood streamed from a wide gash in my right leg. I later assumed that my leg must have banged against the closed half of the cargo door.

Below were sky-scraper trees, jungle grass and masses of intertwining vines. Luckily I was finally caught up two feet from the earth, a simple turn and I was safe on the ground.

My leg needed immediate attention. Fortunately the crew chief appeared from a nearby thicket. He had gone halfway before realizing there were natives on a foraging foray.

Upper Burma was the home of several fierce head hunting tribes. These people, however, proved to be friendly, particularly after we passed out the American cigarettes, a much desired product.

After a relaxing smoke, followed by a round table discussion, the tribesmen used sign language to motion I follow them. We soon stumbled on a small clearing and here was a small lean-to. One native remained behind with the stranded crew member and the rest preceded toward their village.

Two days passed in the lean-to before the first group returned with a homemade litter on which they carried me to a more permanent abode on the outskirts of their jungle hamlet. These Naga hillsman fed me eggs, cracked corn, chicken and boiled rice, along with large thin pancakes made with an ersatz flour.

I was with the natives 19 days and many incidents we shared were on the humorous side. A local witch doctor showed a great desire to practice on my injured leg and I had to use much diplomacy to dissuade the Naga medic yet mange to retain his friendship. Another incident involved what I later figured out to be a pipe full of opium brought to me by the local chieftain. One puff and I immediately recognized the pipe's contents and feigned sickness.

I eventually was reached by the ATC search and rescue unit who had parachuted in to help me. My leg was in bad shape and gangrene had set in, but the original treatment had tempered the infection. The G.I. angels of mercy wasted no time apply the latest wonders of modern medicine. After a short convalescent period I was ready for evacuation. A tiny landing strip was built in a nice clearing, big enough for an L-5 to land and take-off. And 19 days after bailout two L-5's landed and flew me to Upper Assam to the 234th General Hospital.

The Naga hillsmen were rewarded for saving the life of another American with 500 pounds of rice dropped from the air to the villagers. My squadron contributed another hundred pounds each of rice and salt, two staples highly prized by the primitive people. Source: A Sept. 13, 1945 article of "The CBI Roundup" by W.E. Chilton

## MEMORIES OF JOHN TOPOLSKI, RADAR OPERATOR

We made it our photo target and were returning from the mission. Someone spotted a single Jap plane off in the distance. With our guns and the CFC system, we were confident we could defend our plane. We opened up on the plane when it came within range in a head-on approach. Nothing happened to the Jap plane and we were incredulous. The Jap plane hit the nose section of our plane with its guns and was successful with its second attack as well. I could hear everything that was said in the plane because I kept my headphones on throughout.

While dealing with the fire in the forward compartment and treating Lt. Kintis, Sgt. Sandrick (LG) reported trouble with the No. 2 engine. We got orders to jettison everything in the plane. It was hard to see everything we thought valuable a moment ago, being thrown out to try to save the plane.

I was one of those who went into the back bomb bay to try to kick out the auxiliary tank that was hanging half in and half out. No chutes. I don't know how we had the daring to do what we did.

After the wing showed fire, Sgt. Sandrick kept advising the pilot the state of this fire and encouraging him to consider bail out.

When we bailed out, I saw all eleven chutes in the air (in addition to my own). We were told to bail out in as rapid succession as we would so we would land as close together as possible. Lt. Bales, who had stayed behind to make sure everyone was out of the plane was the last to have his chute open. He landed some distance away. The instant my chute opened, I saw the plane's left wing break off. When that happened, the body of the plane nosed straight for the water. It hit the water before we did. It is incredible, when you think back on it, that we were making judgments like we did when we were only 19-20 year old kids. I know I was in shock afterwards and I know it lasted for a very long time. Even though I had been on 16 previous missions, I don't know if I will ever get over the shock of this experience and I don't think others in our crew ever will either.

I was in the water form bailout about 2:00 p.m. on the 26th until about 5:00 p.m. on the 28th.

I didn't release my chute early, before hitting the water, as we were trained. When I landed in the water, I looked up to see this giant chute above me. Somehow I got free of it.

First thing I did upon landing in the water was get rid of my shoes and gun and anything that would weigh me down.

It was an unbelievable relief to see the plane following us see us in the water. They dropped all of their ditching equipment to us, including the small dinghy. Dimock and two other crew members were together. Dimock was able to swim to the dinghy. We couldn't see one another in the water so it was agreed we would call out and Dimock would paddle to the sound. By this means Dimock was able to get together with the other crew members but not me.

A Mae West does not keep you afloat. You have to keep fighting to stay afloat every minute. You can't let your legs drag or you will go under. I got bumped by sharks several times. You bob like a cork on a fishing line. You have to keep the back of your head to the waves and you have to fight constantly to do this. If the water is said to be calm during the day, the swells are really strong at night. You can fell all of the big swells building up as much as an hour before they arrive to hit you.

One of the PBY search planes flew around the area looking one way and then banking to look the other and trying to see directly below the plane. It seemed like the plane was within 50 feet of me. I waved and tried to signal to them, but they didn't see me and eventually flew away.

The desperateness of being alone in the water can't be described. You search in vain for something–anything–to hold on to, even a pencil or a twig, but there is nothing.

The worst was seeing the B-29 making its last circle. You knew they could not stay any longer, but they never saw me. I said to myself, "I'll never last another night." I know that you can never give up or you're lost, but I was ready to give up when this giant submarine surfaced directly in front of me. The submarine crew threw me a line. I resolved that I was never going to let go of that line. I began swinging around toward the stern of the sub and there was concern that I would be sucked into the propellers. One of the sub crewmen, a champion swimmer, dove in and pulled me away.

I can't say enough for the British. They treated us like kings. Years after the war, I wrote the British Navy Information Office, trying to locate members of the submarine crew. The best they could do was offer me access to the files if I should come there to look up the information myself, or if I should hire someone to do the research for me. I was never able to do that. With respect to further details of the rescue, I can only remember being picked up. I have some vague recollection of being in a bunk. I hardly remember being transferred from the sub to the PBY.

The danger to the sub from enemy surface ships was such that after it picked up the first three crew members, the sub was for getting out of there, but my rescued buddies pressed the sub commander to stay around and search some more, because, they insisted, there were more men to be rescued and that is the reason they found me.

I don't remember anything about the return to Calcutta or the hospital. I know I promised myself that I would never go over water again. So what happened? We were taken up for a test flight and certified by the flight surgeon for return to flight status. By the time we were released from the hospital, the group, of course, had moved to Tinian. So we went by boat to Tinian. Thirty days on the water. Shortly after arriving at Tinian, we were detailed to Iwo Jima to fly escort ship for P-51's. We flew virtually every day. Our last mission was August 15. Among my duties as radar man was to locate air-sea rescue submarines along our route. You can be sure I located every one of those subs every trip. Source: 40th Bomb Group Association Memories, May 1988

## MY ONE AND ONLY

*by Eugene A. Wink, Jr.*

Twenty-nine aircraft of my P-47 fighter group, the 365th, took off from our base at Gossfield, England at 1034 on March 2, 1944. Our mission was to provide fighter protection for the 8th Air Force B-17's on a bombing mission in German. Twenty minutes later rendezvous was made over Bastogne, Belgium. Suddenly 14 yellow-nosed Focke Wulfe 190's dov from 27,000 feet on the bombers which were flying at 25,00 ft. The dogfight that followed covered over 40 miles of sk The results were six FW-190's shot down and the loss of or P-47 - my aircraft.

I was flying as element leader in the lead flight of fou After the initial encounter, the enemy scattered, and w spotted a single FW-190 5,000 feet below our position. As th flight leader led the attack, the German pilot saw us comin and headed for the deck through the clouds in a hurry. W closed rapidly because of our diving speed, and withi seconds, the German crashed in flames in a French forest, a of us having had an opportunity to fire on him.

Our escort duties finished, we started our climb back t altitude so we could safely cross the English Channel t "merry ole England." My troubles began at 11,000 ft. whe my engine began losing power. Notifying my wingman c my critical situation, I went through all of the procedures t determine the cause of the problem but with not success. A 13,000 ft. my engine simple quit. I asked my wingman t notify my Dad, who was also in England, and tell him tha with a little good fortune I hoped to see him in about si weeks.

Decision time was at hand - bail out or crash land? Sinc the latter would provide the enemy with a flyable P-47, elected to bail out. All I had to do now was to decide how t go about it. Since the P-47 glides like a rock, my altitud advantage and time were rapidly vanishing. I scratche going over the side of the cockpit when I recalled hearin about fighter pilots having their parachutes caught on the tai section. The only thing left for me to do was to turn th aircraft over and fall out.

Having made the "exit" decision, I began to go throug the detailed procedures required to make my bailout a re sounding success. With my time rapidly expiring, I checkec my parachute straps, disconnected my headset and throa mike, buckled my helmet chin strap, put my goggles over m eyes, and opened the cockpit canopy. An important next step was to roll my elevator trim tabs forward. This would hel the aircraft continue its forward movement when inverted rather than flying toward me.

By this time I passed through 3,000 ft. and felt a very strong urge to hit the silk, so I rolled the aircraft and released my safety belt. The next few moments were absolute exhilaration. The jolt of the air that hit my face as I left the cockpit almost took my breath away. I knew I was free of the aircraft, and I had a feeling of being completely motionless in the air as I looked the world over. I was experiencing an emotional joy because things were working right and I was safe so far and because of this new and unique experience There was no tumbling, only a good feeling, and the question now on my mind was "When do I pull my ripcord?" To shorten my float time because of the possibility of enemy troops in the area, I delayed opening the chute a few seconds.

My one and only parachute jump was an experience shall never forget. The thing most indelible in my memory i that the descent was marked by complete silence. I had very little time to experiment with the feel of the risers because .

was mesmerized by the experience of floating down in silence. As I approached the ground, I was jolted back to reality by narrowly missing telephone wires, and I hit the ground with a thunderous thud. I tried to follow the procedures drilled into us by our parachute experts – feet close together, knees slightly bent, and rolling with the fall. The result was the landing was safe and without injury but completely lacking in grace and style.

After collecting my parachute and my wits and getting the dirt out of my mouth, I saw a truck loaded with German soldiers coming over a hill in my direction. This was the beginning of a long journey and many exciting and harrowing experiences: successfully evading the enemy throughout France, across the Pyrenees Mountains, and into Spain. And, yes, I saw my Dad six weeks from the day I qualified for the Caterpillars. Copyright 1990 Eugene A. Wink, Jr. Used by permission.

## CRASH LANDING
### by Clyde Snodgrass

Our C-82 "Flying Boxcar" was flying on a return trip from MacDill field in Tampa, FL to our home base at Maxwell field in Montgomery, AL. One of the two engines failed shortly after passing Dale Mabry airfield and our pilot, Lt. Lawrence F. Tapper, circled to return for a landing. En route, the other engine developed trouble. When both engines went dead and the plane began to lose altitude rapidly, we were ordered to bail out.

The pilot, co-pilot Lt. Robert B. Parker and radio engineer Sgt. James Wood elected to stay with the ship to try and land it. Five of us bailed out. Only one suffered a severe injury. A passenger, Navy Lt. Bowen Larkin, fractured his ankle when his parachute landed him in a tree.

The plane crash-landed in an open field, bumped its way uphill on its low-slung belly, clipped a guy wire but missed a power line and skidded to a halt near a planted pine forest, resulting in only slight damage to the aircraft. Scores of local residents rushed to the scene and many helped search for the five of us who had jumped and landed over a wide area. Source: *Alabama Journal*, Montgomery, AL Tuesday Aug. 1, 1950.

## BUSH'S WAR
### condensed from the story by Ernest B. Furgurson

George Bush, President of the United States, was a clean cut, athletic young man. Fascinated by airplanes and the Navy, he wished to be the first in his town to become a naval aviator. He was anxious to join the service after the attack on Pearl Harbor, but his parents persuaded him to wait until after graduation to enlist. On June 12, 1942, Bush's 18th birthday, he was sworn in at Boston, MA as a seaman second class. This was the traditional path to becoming an aviation cadet, his reason for joining the Navy. Bush had never flown before, but went to Chapel Hill, NC to pre-flight school along with fellow cadet Ted Williams, the Red Sox slugger, who later became a Marine fighter pilot.

The Navy needed more pilots than it had, so Bush's class was rushed through pre-flight school and then sent to Wold Chamberlain Naval Air Station in Minneapolis for primary flight training. There he learned to fly in the PT-17, the Stearman with an open cockpit. Cadets were forced to wear masks against the bitter Minnesota cold. Bush was glad to pass that course and head for the warmth of Corpus Christi, TX. "Until I got there I know I never landed on a runway without snow or ice," he remembers.

From basic he went on to advance training in the AT-6 Texan, but in the Navy they called it the SNJ. In early June 1943, Bush earned both his wings and his gold ensign's bars. He was assigned to torpedo bombers, and his ship was the USS *San Jacinto*. After shakedown cruise to Trinidad, she sailed off to the war in the Pacific. *San Jacinto* headed west from Pearl Harbor on May 3, 1944 to join Task Force 58. She was top heavy, thin skinned and lightly armored, but fast. She could carry 34 planes and run at a top speed of 34 knots.

Ensign George Bush was aboard as a pilot on a Grumman Avenger, a TMB, the largest carrier based plane the Navy had, and behind him, sat the rear gunner, and below him, in the rear, was the radio man. His first combat was a raid on Wake Island, where U.S. Marines had created a legend in fighting off the enemy fleet, in the dark days shortly after Pearl Harbor. As the American pilot flew low-level cover through heavy anti-aircraft fire for landings at Guam and Saipan. At that time, he didn't know that his roommate would fail to return.

His big test came when the Japanese struck during the battle for the Marianas. More than 300 enemy aircraft attacked the American fleet on June 19 in what the ship's log called "One of the biggest air battles of the war." Bush and all of the other pilots scrambled above to engage the incoming aircraft and also to protect their planes. Poised on the catapult, Bush took a final look at his instruments and realized that he had no oil pressure. It was too late to abort his launch. After he was aloft for a few minutes, his engine sputtered and went dead. "I had to land the damn thing in the sea. That wasn't a big, heroic thing to do, but I had never done it before." As he set the big plane down in the calm sea, Bush and his crew had hardly gotten their feet wet when they were picked up by a destroyer. For Bush this was only a faint foreshadowing of the coming battle, one that would earn him the Distinguished Flying Cross.

On September 1, Squadron VT-51 hit Japanese radio stations on Chichi-Jima in the Bonin Islands. After the bombing, they found the radio stations still functioning. VT-51 would have to return the next day. On September 2, four Avengers from the San Jacinto, together with eight Helldivers, and a dozen Hellcat fighters, took off at 7:15 a.m. Each of the Avengers carried four 500-pound bombs.

The defensive crossfire at Chichi-Jima was intense. Don Melvin, VT-51 Squadron commander, led the first pair of bombers in. They destroyed a radio tower and damaged a lot of buildings. George Bush then came in with Milt Moore following. By this time, the enemy fire was focused on the Avengers as they approached their target. Bush nosed over into a 30 degree glide, straight on course. Then all hell broke loose. Bush's engine was hit by anti-aircraft fire. "Smoke started to pour out of the damn thing. It's hard to remember all the details. I looked for the instruments, but there was too much smoke in there to see the damn things, and we were going down fast, and as I pulled out over the island, I realized that I was in big trouble." Smoke pouring out of his engine, flames sweeping back along his wings, Bush continued his bomb run. After heading back to sea, Bush leveled the plane to give his crew time to bail-out. Then he called, "Chutes." When Bush himself plunged over the side, he hit his head on the tail and jerked his parachute ripcord too soon. The chute caught on the tail, but by pure luck it tore free. He was falling too fast, although stunned by the blow to his head, he managed to slip on his harness before it hit the water. The parachute blew off toward the island. His seat-pack life raft had fallen free. Then a Hellcat swooped down and drew his attention to the raft. He swam for it. After pulling himself aboard, his first reflex was to check his pistol. "I pulled that damn thing out to see if it was working or not. I didn't know what in hell I was going to do with it." He was alone in the ocean. He grieved and wondered what had happened to his crew. "It seemed just like the end of the world," he recalled. He wondered whether or not he or his men would ever be found again.

He paddled steadily for over two hours to keep his raft away from the enemy held island. An then barely 100 yards away, out of the depths, poked a periscope, followed by a shiny black American submarine, the USS *Finback*. Within minutes he was aboard and the ship slid silently back into the ocean.

The citation awarded with his Distinguished Flying Cross was worded as follows: Opposed by intense anti-aircraft fire, his plane hit, and set afire as he commenced his dive. In spite of smoke and flames from the fire in his plane, George Bush did continue his dive and scored damaging bomb hits on the radio station before bailing out of his aircraft. His courage, and complete disregard for his own safety, both in his attack in the face of intense anti-aircraft fire, and continuing in his dive on the target after being hit, and on fire, were at all times in keeping with the highest traditions of the U.S. Naval Service.

# A Hole In The Prison Wall
*by Dale V. Lee*

Our last trip was the first Foggia mission. We were hit after we released our bombs. Our plane was on fire - the fire following the transfer hoses turning the airplane into one big inferno. My clothes were burning off and our communication system was gone. There was no way of knowing what was going on up front, but I knew it was time to do something. I poked Joe Worth and pointed to Hickerson in the tail. I bailed out the left waist window.

We were approximately 25,000 ft. and the cold air felt oh so good. I free-fell as far as I thought I could and about that time I fell through a bunch of gun fighters and recalled an image of a man shot in his chute. I delayed opening my chute and remember floating in what seemed a slow-turn. I watched as our burning plane went down and though it seemed like still a long way to the ground, pulled my chute. I felt first one jolt and then another not realizing at first that I was on the ground. I was immediately surrounded by Italian civilians with weapons.

Our navigator had a ripped chute and our bombardier was killed in the plane. The rest of the crew made it safely to the ground.

I was taken to a civilian jail where there was at least si[x] inches of human excrement on the tiled floor in the two cells There was another soldier there who had a broken right arm with the bones protruding about six to eight inches. The man was in extreme shock but could not lay down due to the filth I managed to take the door between the cell and lean i[t] against the wall at a slight angle so he could lay down. H[e] passed out but was still breathing the next morning when th[e] guards came to get me.

I was taken to Beri for interrogation, then was parade[d] through the streets. Absolutely no compassion was show[n] for the guys with broken bones, severe burns and wounds Some had eyes so swollen and bruised they couldn't see. Th[e] civilians were throwing stones at us, spitting on us an[d] shouting jeers of contempt.

From Beri I was moved by train to Selmonia, Italy t[o] concentration camp number 17. Here I found 20 American[s] and over 2,500 British prisoners, many had been physicall[y] and mentally abused and had inadequate food.

We knew we had to try to make a "break". The camp wa[s] surrounded by an eight foot cement and block wall wit[h] jagged glass pieces imbedded on top side. Outside the wa[ll] was a 20 foot wide road and beyond that there were electri[c] high tension wires. Of course there were guard towers manne[d] 24 hours a day. Escape would not be simple.

One day some important Nazi officers arrived. Ther[e] was a big commotion at the main gate and the Italian guard[s] left their towers to investigate the excitement. This was ou[r] chance. One of our English buddies had worked in the powe[r] house so he knew that the power was off during the day. I[n] one section of the cement wall was a small section of bric[k] blocks, so with our crude tools I broke a hole in the wall Somehow we got through the power lines and barbed en-tanglement and then we ran like hell.

We ran as far as we could on that first night, then sli[d] down off the slope of a steep mountain road catching brush trees or whatever we could grab. We straddled the brush an[d] leaned back up against the bank in a sitting position and were able to get a little sleep to renew our energy. Military patrol[s] searched the area for us all night, even the road only 50 f[t] above us.

There were six guys in our group: Ray Whilby, Joe Jet[t] Tom Percell, Wesley Zimmerman, an Englishman and me Our plans were to head for the high mountains as ou[r] chances of being seen would be in less populated areas. W[e] would "observe" and plan by day and walk during the night We used the North Star as navigator and headed for the boo[t] of Italy as we figured the Allied invasion would be comin[g] from that direction. We survived mostly on what we coul[d] steal - figs, grapes and garden stuff. Occasionally we wer[e] able to bargain for good. Joe Jett traded his jacket for a hun[k] of cheese. It tasted so good - that is until the next morning and in the light we could see it was full of worms, big fat worms We weren't about to throw it away so every time a worm crawled out we just eliminated the rascal with a quick flick o[f] the fingers.

I had a lot of bad sores on my legs - probably from shrapnel wound - that were infected and would not heal. Jo[e] Jett said he had heard that garlic was a good blood purifier. We found garlic in a garden and I ate three big cloves. We

*Crew of Southern Comfort: L to R, bottom: Findor, A.T. Fabiny, H.W. Austin and P. Singer. Top: J. Warth, Dale Lee, G. Hickelson, E.L. Shaw, J.W.B. Jett. (Courtesy of Dale Lee)*

usually walked single file and I was the lead man. After my "garlic feast" the guys made me walk in the rear so they wouldn't have to smell my garlic breath.

One night we stopped at a well-kept farmhouse that looked quite prosperous. "Ma-Ma" was cooking spaghetti and it smelled so good until I noticed that "Pa-Pa" was not around and I hadn't noticed when he left. Instinct made me suspicious and I told the guys that I was getting the hell out of there. It was hard to leave as we were really hungry but we all did. After getting to a hiding place up the mountains we saw "Pa-Pa" returning with German soldiers.

We had a number of close calls. One night we came to a high railroad bridge. We had debated whether we would walk under the bridge or retrace our steps and walk around it. It would mean approximately 10-15 more miles. We watched the bridge for a day and had not seen any guards so we decided to take our chances and walk under it. We were all tense and I was wondering if I'd hear the "shot that killed me". As we got close to the bridge one of the guys said, "Hey you got to stop. I gotta take a leak!" Joe Jett said, "Well piss in the your pants! You've been wet for a week." Jett had always been such a gentleman, but when you've been cold and hungry and tired and walking in the dark and rain wondering if you're going to be shot in the back, it's easy to understand losing your cool.

Another time we came to a railroad crossing - we had been skirting this track for two or three days and there had been no trains - so we were about to make the crossing and were surprised to head a train whistle, sort of a quick tweet-tweet. A train was coming around the corner and we all dropped to the train bed bank. That train was loaded with German troops and it stopped about 50 ft. in front of us, unloaded two civilians and took off again. When all was quiet again, we made a "quick" crossing.

Sometimes we misjudged suitable hiding places. Once we had spotted an olive grove that looked like a great spot to hide. When we neared the area we discovered it was an expertly camouflaged German motorized equipment and troop camp. Needless to say we made a silent retreat!

We spent five days in some bush on a steep hillside. There was an exchange of artillery from east to west and vise-versa and we didn't know who was firing on whom. We knew of a small village above us and felt that perhaps we could get something to eat. There was a high wall around the city and up to the heart of the village. We followed the wall with our backs flat against it until we were directly under the bell tower of a church. It was a really dark night and the village looked deserted with not a light or a person in sight, not even a stray dog or cat. Instinct again told us that it was

17

This is the hole Dale Lee kicked in the prison wall at Selmonia, Italy before his escape. (Courtesy of Dale Lee)

A monument at the main gate at Selmonia, Italy. It is in memory of the Italian prison camp commander who was executed by the Nazis. (Courtesy of Dale Lee)

too quiet so we returned to our hiding place still hungry. We felt trapped and became even more jumpy.

The next night there was a ruckus in the village above us. Soon three German soldiers came running from the village directly towards our hiding place. They were very excited and carried side arms and parts of a machine gun and tripod. They set up about 100 ft. from us, deliberated a few minutes, then quickly moved further down the valley. We got the feeling they were running away from something so we went up to the village again to investigate. We learned that our receding German soldiers had been stationed in the church bell tower the night before, which explained the quietness of the village.

A jeep manned by two Canadians (with the British 8th Army) arrived and we approached them and had a most welcome chat! They told us to "pile in" and they would take us to their camp. To get there we would have to climb a steep slope covered with six to eight inches of mud. We held on as the four-wheel drive Jeep crept slowly forward spewing mud. Meanwhile, the Germans were bobbing shells at us but they landed behind us. I wondered if the irony of my surviving bailout, prison camp, and all the other hardships would be that I'd "get it" on this last hill.

We made it to camp and were ever so happy to be back. There was a lot of commaradie with our new-found friends.

The Canadian lieutenant (driver of the Jeep) insisted on taking us to a "dry bed with sheets". Another three hour drive in the rain in an open Jeep and we were in a sort of Red Cross setup. We even had a dry cot to sleep on. It was pure luxury!

The next day we were taken to the nearest American camp, which was the 47th F.G. (flying P-40s). We tried to convince them to go and blast that alive grave but they didn't believe us until one day when one of the fighters happened to have one of his bombs hang up over target. On his way back to base he swung by the alive grave. He loped in his bomb and stirred up a hornets nest. He came back to base to report his find and the 47th went back in force and had a "hey-day"!

We spent several days with the 101st Airborne. As we were going through the mess line one day I went back for a second helping. The food was being served by Italian POW's and a certain POW refused to give me a second helping and

I just blew! I grabbed him around the neck and we went round and round, pots and pans all over the place, as the guys from the 101st kept cheering me on. Great guys, that 101st!

We got a flight from there to Africa, close to Tunis. We had a difficult time convincing anyone of our identity and our story. We kept bugging a certain major, trying to convince him. He finally became irritated with us and said, "Well I've got you now! Anyone on the Polesti raid is to be awarded the Distinguished Flying Cross and I've got the list!" We said, "Dig it out!" He got the list and our names were on it and he decided that maybe we were speaking the truth.

I was decided to have us decorated by General Doolittle at the 12th Air Force headquarters. We were to have our pictures taken and prior to that we were to get five minutes in Doolittle's office to meet his personally. It turned out to be a 45 minute to an hour session. We asked him about his Tokyo raid and he replied, "That was nothing compared to your raid on Polesti" and he wanted to know all about it - our prison camp experience, our escape and that sort of thing. We had a very easy, friendly conversation.

Gen. Doolittle wrote our return orders - back to England and the States - a good set of orders marked "SECRET" and signed by the general himself. We left there and stayed at an air base near Tunis and then caught a flight back to England.

## ENGINE FAILURE OVER TARLAC

*by Frank F. Marvin*

At 0735 on 15 March 1948, I took off in number two position in a F-47N number 488550 of a four ship flight. Numbers three and four did not take off for various reasons which left Lt. McGuff and myself a flight of two on an individual combat mission. We had been thoroughly briefed to conduct four individual fights climbing to 20,000 feet and fighting down to not below 8,000 feet and climbing back up to 20,000 feet. Also, we were individually instructed to use not more than 50" manifold pressure nor more than 2,600 RPM.

Prior to take-off, I made the usual complete check. Everything check perfectly. I made a normal take-off and we climbed in close formation to 20,000 ft. we then began our initial fight during which there were no unusually difficult maneuvers performed. Steep turns, dives and moderate pull ups never exceeded four-and-a-half G's. At no time did I use

nore than 46" or 2,600 RPM. When we reached an altitude of
,000 ft., we rejoined formation and began climbing back to
0,000 ft.

At approximately 18,000 ft., my engine stopped abruptly
nd the propeller windmilled. I immediately lowered the
1ose and held an air speed of approximately 150 mph. I called
_t. McGuff and told him of my emergency, at the same time
performing emergency procedures of checking my mixture
ontrol, which was full forward, switching gas tanks from
nain to auxiliary and turning on my emergency fuel booster
pump. At that moment I smelled excessive gas fumes in the
ockpit so immediately opened the canopy, turned off my
emergency booster pump and cut my mixture. I did not turn
off my fuel selector, electrical switches or turbo.

By this time, I was at an altitude of approximately 13,000
t. and was headed directly for Clark Field at a heading of 190,
ibout five miles east of Tarlac. As the engine was still
windmilling and my fuel pressure was 0, I knew I had
complete engine failure and was preparing for an emergency
anding or bailout. At that time there was an instantaneous
explosion which engulfed me in flames and terrific heat. I
was conscious and immediately released my safety belt and
bailed out the right side. I did not release the canopy, cut the
switches or unfasten any of my equipment.

The bailout was completely successful. I estimate my
chute opened at approximately 10,000 ft. As I parachuted to
the ground, I could see in the sky a large ball of small pieces
of floating in the air and assumed that the aircraft had
disintegrated. My descent took approximately ten minutes
and I landed in a large open plowed field. The time was 0835.
I immediately took off my harness, Mae West, and other
equipment and removed my undershirt to sue as a handker-
chief of bandage as I was bleeding profusely from a head
wound.

There were numerous natives in the vicinity but none
came out after me for several minutes. Lt. McGuff had circled
me all during my descent and bussed me after I was on the
ground at which time I waved to indicate that I was alright.
I then walked to the edge of the field and the natives escorted
me to the M.P.C. who took me to the Philippine army hospital
at Tarlac. I was treated there and waited until I was picked up
by Clark Air Force Base personnel.

## DEAR FOLKS

*A 1945 Letter by Glenn Miller*

I'll now try to explain my experiences of the last few days. I
took off here (Kurmitola, India) on the morning of the 12th
(March). We could not see, and got off the runway on both
sides before we finally did get off. Well, I thought those
superchargers were the greatest inventions. Now I think the
parachute is. We got to China and delivered our gas and I had
a little trouble with the carburetor. By the time I fixed it and
we got off the ground, it was 1835 hours (6:35 P.M.). We took
off and climbed to 22,000 ft. Everything was OK until we hit
a storm. Then the wind, rain and snow started to play hell
with us. It threw us around like a feather. Our compass
wouldn't work because of the electrical storm. Lightening
was dancing on our nose. We called our field and told them
we were five minutes away. They told us the weather was

clear there. We flew on looking for our field and never did see
it. We got lost. The wind was blowing us around and there
was ice. The gas was almost gone, and it was dark as the ace
of spades. I knew then we would have to leave our ship. The
pilot said "strap on your chutes". We did, and I was supposed
to jump first. I said "good luck" and crawled into the bomb
bay, opened the doors, and waited for the rest of the crew.
Funny, I wasn't scared as much as I thought I would be. We
were up 19,500 ft. and I was too busy keeping the engines
running to use oxygen. I was a little groggy and couldn't
think straight without oxygen. I wanted to keep my flashlight
but couldn't think of where to put it. I took my hat off and put
it in my pocket. The radio operator crawled down behind me
and the co-pilot behind him. I threw my flashlight and
jumped out after it. I knew I was high, and wanted to drop
below the storm before opening my chute. I was afraid it
would collapse in the storm. I waited and dropped. Then I
figured I may as well find out if it will open. I pulled the
ripcord and "crack"!. I was falling about 100 mph when the
chute opened. I stopped dead. It was a helluva jar. It pushed
my stomach up and the leg straps were tight around my legs.
I grabbed the shroud lines and pulled myself up. I could
breath better and felt better. I let myself sag in the harness and
thanked God for opening it. I felt as if I'd vomit and pass out.
Then I figured if I'd land in water I'd want to stay awake. I fell
through snow, rain and wind all the time. I forced myself to
stay awake. I was swinging back and forth. The lightening
flashed and I could not see the ground. I said The Lord's
Prayer and started to sing to pass the time. Later the lightening flashed and I saw a tree right below me. It was a pretty
sight. It meant solid ground below. And it was solid. I hit soon
after. As I was swinging, I hit on the back stroke. My feet hit
first, I went down on my back, and my head hit. Not hard. It
took me 25 minutes to hit the ground from the time I left the
ship at 2320 hours. I landed in a plowed field and it was so
dark I could just about see the trees there. I unbuckled my
chute and lay there to get my bearings. Every time the
lightning flashed, I looked around. Couldn't see any kind of
civilization. I didn't know where I was, so I decided to wait
until morning to try to find someone. I stretched my chute
between two trees and slept on it as a hammock. Next
morning I woke up and looked around. Couldn't see anything but trees and flat land. Then I saw a WOG in a two-
wheeled cart drawn by a horse. I called to him, but he beat the
horse and ran away. I picked up my chute and jungle kit and
followed his tracks. I came to a farm and got a WOG to talk
to. He took me to the next farm, where I met about 30 more
WOGS. They brought me a chair to sit on, and I tried to tell
them what happened. They did not believe I came out of an
airplane. When I mentioned Rupees, the one said he would
guide me to civilization. I gave him 100 Rupees after he led
me to a road about eight miles away. All I had was two 100
Ruppee notes. I'll get my money back. I walked along the
road, then saw a radio station off to my left. I walked across
rice paddies to the station. Our radio operator came from the
other direction. We went into the station, taken care of by
Indian soldiers, and sent a message to Calcutta saying we
were safe and told them we would wait for further instruc-
tions. Later a truck brought in the co-pilot and the three of us
had a good time there. Played cards, did a little hunting and

looked at India. We landed near Calcutta. That afternoon, the Indian soldiers took us to an R.A.F. base about 30 miles away in a truck. The Indians at the radio station treated us nice. They could speak English, and gave us hard boiled eggs and rice patties. The British gave us a meal and a bed for the second night. Next morning, we went to their air field. About an hour later, a C-47 from our field landed there. They were looking for us, and had to land there to fix a loose gas cap. They brought us back here at 1130 on the 14th. We saw our ship on the way back, and buzzed it a few times. The pilot got back here the day before we did. Our ship is laying in a field 120 miles from here, upside down, and the tail and wings are off. A lot of my equipment is in it. My gun and knife, too. I'll get a re-issue. We told our story to Intelligence today and straightened things out. I think I will get two weeks rest leave at Calcutta. I'm OK and never felt better. Had a nice experience, and an now a member of the "Caterpillar Club", because a parachute saved my life. Maybe it was the silk and nylon from the stockings you didn't get last year. Well, all OK, so I'll try to answer your letters. I received your box with the shorts and cigarette lighter. Also a letter from Ann, Bud's wife. A fellow from Kunming gave me a package to give to a fellow here. It's still in the ship. They sent a "missing" telegram to you. Hope you did not get it. We sent a chaser after it. Well, I'll hit the sack now. Hope my next letter is from a rest camp. I don't need a rest, but they give you one, anyway.

Love, Glenn

## BAILOUT BEHIND ENEMY LINES
### by Paul W. Pifer

I became a Caterpillar Jan. 1, 1945. I was a radio operator-gunner on a B-24 Liberator bomber. Our ten member crew was on a mission over Germany. We were shot down over Trier, Germany and forced to bail out of the ship.

Four of the parachutes didn't open, including the one I was wearing, probably due to the extreme high altitude of the jump. Each of us tried to open the chute by hand. The four of us were pulling out the parachutes from the front of the packs with our bare hands. Three of the four were successful. One chute never opened and we lost that crew member.

We landed about six miles behind enemy lines in a hostile area of France. We were help by Partisans (Free-Frenchmen) who lived in the hills. They picked us up and saw to it we got to safety. It was scary as hell. We didn't even know where we were. I had broken my pelvic bone in the jump and the Partisans sent me to a hospital to recover from my injury. I was a smart-alec kid and had not seen to be that my parachute straps were tight enough.

I spent two weeks in Vittel, France and then returned to the war to finish my tour. I flew 33 missions over Germany.

On another mission the plane was hit and I was almost forced to bail out again. Fortunately the crew was able to make an emergency landing at a fighter base in France.

A third scare came 600 miles off the coast of Spain. Again bailout was avoided. The crew managed to dump all the excess weight off the plane and we made an emergency landing in a sheep pasture.

*Pifer is the national surgeon for the Caterpillar Associa-

tion and had the honor of reading a special letter from another Caterpillar, President George Bush. The President was answering Pifer's invitation to attend a group reunion in Colorado Springs, CO. Although Bush was unable to attend the reunion, he wrote a personal letter expressing his pride in Caterpillar membership.

Source: Aug. 6, 1989 St. *Tammany News-Banner* feature by Michelle Kropog

## SEVEN HOURS A PRISONER
### by John F. Rauth

We took off at 0600 hours on Friday, July 23rd, on a fighter escort mission over a small island off the coast of Sicily. After the fighters we were escorting had dropped their bombs, our top cover flight went down to strafe radio installations and targets of opportunity along the coast. We hit radio installations, a small boat and a lighthouse, and as we left those we sighted two speed boats about five miles offshore. My section leader and his wing man went down on the two boats first and then my wing man and I went down and strafed.

Just as I passed the boat, I got a smell of coolant and immediately my engine temperatures started to go up. I climbed up to about 800 ft. and throttled back to see if I could cool the engine off, but it seemed impossible to do this. So I bailed out some place below 700 ft. I went over the right side and under the horizontal stabilizer. As soon as I was clear of the airplane, I opened the chute and before I could completely get out of the parachute, I hit the water. My airplane had hit the water before my parachute opened. The wind carried my parachute and dragged me like a surfboard for quite some distance. Finally, I was able to spill the chute and in doing so I became entangled in the shroud lines. However, I was able to remain afloat with my Mae West and got my pocket knife out and cut myself free of the parachute. Then I inflated my dinghy. The dinghy upset once when I tried to climb onto it, and I was in the water about 15 minutes before I finally got into the dinghy. It was now about 0720 hours.

I thought of everything down there, and I decided that at first I would paddle to a lighthouse, which I saw about three miles away; and then I changed my mind and decided to stay adrift at sea until dark so that I could come to shore and hide myself until the Americans took the island. The wind was drifting me paroled to shore at about the rate of 15 mph. Flight officer Bill Slattery of 1201 S. 10th Ave., Birmingham, AL, my roommate in flying school, was circling above me and it made me feel very comfortable to have him around. He then came down on the water low and started to fire his guns. I turned around and saw that he was firing in front of a fishing boat. I did not know whether I wanted him to fire or not, but was helpless to do anything. He tried to keep the boat from picking me up because he knew that the Air Sea Rescue was on its way. When his gas got low, he had to leave and the fisherman in the boat picked me up. There were about six fishermen in this boat. It was about 0900 hours then.

The fishermen seemed very glad to see that I was alive and not hurt, and hugged me and kissed me like a long lost brother. They had some dry trousers and some melon for me. Then we started back to shore, rowing against the wind. As we neared the shore, I could see my friends coming back in

P-40s to search for me. By then it was 1130 hours. The Italian soldiers on the coast made us bring the boat into a little quay, where they wanted me to surrender. They had seen me bail out of the plane.

Then this one boy in the boat was going to surrender himself as the American pilot. But the Italian soldiers detected that he was a native, so they became very rough with him and as far as I could tell were going to kill the boy. I had been hiding in the boat while this was going on. When I saw the bayonets and saw them getting tough with him, I decided to give myself up. I had on these civilian trousers, and I rolled up my wet clothes and crawled up on the rocks at the point of the Italian guns. That was rough - all those Italian guns pointed at me. I came out with my hands up. It was the first time in my life I had looked down the barrel of a loaded gun.

There was a first-aid corporal there, who swabbed my scratches and bruises with some kind of soothing medicant. Then they took me to an Italian farmhouse and gave me some water to drink, and took away all my personal belongings except my jewelry. We then got on a motorcycle and they took me to a command post, about eight miles north of Trapani. The place was a CP for the pillboxes they had along the shore. Here they gave me some grapes and Italian hardtack, so hard that I couldn't chew it After waiting there for about ten minutes we on a motorcycle again and went to a civilian home in the outskirts of Trapani. Why I was taken there I do not know. The owner of the place, a man about 80 years old, had spent 20 years of his life in Omaha, Nebraska, 120 miles from my home. His wife gave me some bread and grapes and a glass of wine. I refused to drink the wine because I thought that perhaps they wanted me to get intoxicated so that I would talk freely. After spending about 15 minutes here, I was taken to Italian headquarters in the city.

I went to the commandant's office and the interrogation was begun. The interrogator was a lieutenant colonel, Costantino Bruno, who spoke through an interpreter, Captain DiGiovanni Salvatore. The captain had attended Public School 17 in Brooklyn, NY, for seven years. There also were two lieutenants and a major in the room, and an armed guard stood outside the door.

(Everywhere we had gone, soldiers and civilians gathered around with a gazing curiosity at probably the first American pilot they had seen. Some thought I was of Italian decent because I have a pretty good north African sun tan.)

At the interrogation, the captain told me that my name was German and that my ancestors came from western Germany, which is almost true, my great-grandparents all came from Luxemburg. He gave we quite a lecture about coming to fight against my ancestors and to destroy the mother of civilization, which gave us art, literature, architecture and music. He stressed Marconi and said we could not fight the war without the invention of Marconi. He was also emphatic about out not fighting against the Russians. It seems that all the Italians fear Communism.

During this period, I neither approved what they said, nor answered any of their questions, and more or less ignored what they said or did.

They brought me in some food for lunch before the real interrogation began, and while I was eating, they carried on a conversation about me. The food was chicken broth with rice, Italian bread, wine, a little bit of cheese and some kind of jam. I believe that the food that I had was for Italian officers only because I know the rest of the army doesn't eat that well.

After dinner they asked me my name, rank and serial number and a host of other questions which I was not at liberty to answer. I had my pilot's identification and my AGO card. From this, they were quite confused because my pilot's card rated me as a S/Sgt. Pilot as of Sept. 6, 1942, and my AGO card rated my as a flight officer as of Dec. 5, 1942. Finally, I convinced them that my rank was flight officer by the dates on the card. I also had my dogtags and from this they got my home address. They found out my age from my AGO card. They asked many pertinent questions about where my outfit was, what kind of plane I was flying, if I liked it, the number of my outfit and how many crew members were on my plane. But I ignored their questions or told them that I could not answer them.

They asked me if I was married and how many brothers I had and if any of them were in the Army; what my civilian occupation was; how long I had been overseas; if I was tired of fighting the war, and if I was hurt and was I being treated all right. They were definitely sure that I was going to be treated right.

It was apparent to me that they expected to be prisoners very soon. They asked how we treated Italian prisoners, if they had to work, and also if it was true that we were letting the Sicilians go back to civilian life. Of course I couldn't answer any of those questions, except that they would be treated well.

They all became very friendly and were inquisitive as to how the prisoners would be treated by Americans. They told me they thought they would be prisoners in a few days and the commandant said that perhaps he would release me shortly. About that time, a messenger corporal came running into the headquarters babbling away in Italian, something about an "alarme," I took it that the Americans were going to shell the city. So the colonel, the major and a captain took me in an Italian jeep with a driver outside the city and up on the side of a mountain between Erice and San Marco, north of Milo airdrome. On the drive up there, I could see the field artillery plane flying around spotting targets for fire. Before we got to the command post of the pill boxes on the mountains, our paratroopers and airborne infantry began to shell gun positions. I probably could have escaped, but I was afraid that I might be right between the fire, so I figured that the best thing for me to do was to stay in this quarry where the CP was located.

At 18.15 hours the shelling ceased: the admiral had surrendered the city. The enlisted men chucked their rifles and machine guns down in the quarry and the officers destroyed maps and portfolios, gave me their weapons and field glasses and said they were my prisoners. Meanwhile, I was watching for American troops to come into the city. It seemed ages before I saw the first American coming around the mountain road from the east. I had never seen a paratroopers in his jump outfit before, but recognized the American flag on their sleeves by looking through the binoculars. I could see them routing out prisoners from gun emplacements, homes and bivouac areas.

The captain asked that I call an American officer and a

sergeant to take over the prisoners. He also asked that I tell them I had been treated well and that they expected like treatment. In fact, he even wanted to know if he could put on civilian clothes and go back home. That I refused.

I called to the American troops and said, "Hey, soldier, I got some prisoners down here for you."

He called back and said, "Hold them, I will be down there in a minute."

When they came within a short distance of me, they were surprised and astonished to find that I was an American flying officer, because I had no insignia whatsoever, and I had Italian weapons and field glasses. Then I thought I would like to have some souvenirs, so I asked for one pair of binoculars and a gun. The sergeant then told me to stay with them and they would take me to their command post.

We started down the road, prisoners with us, and then the paratroopers decided that since they would be going all over town routing out prisoners, they better let me go by myself to their command post. They gave me directions and told me to take an Italian motorcycle. I took the motorcycle and started out and was stopped about every block by the Americans for identification. I finally reached division head-quarters.

I wanted to send a message to my outfit and was told the message center would be set up in an hour. The speed with which the airborne infantry and paratroopers operate really astonished me. They had marched on foot with full equipment from their jump point which was approximately 175 miles away by road, fighting all the way, and they looked as if they could walk another 175. They had set up a command post and division headquarters an hour and a half after the city surrendered. We even had a mess where officers could eat out of plates.

That night we slept under blankets on the ground. There were no flies or mosquitoes, but the ants were a nuisance. The next morning I tried to arrange for transportation back east on the island to some air transport base. However, there was none available. A colonel in charge of the medical section had found some American and South American airmen in the Fascist military hospital on the side of the mountain. I was inquisitive and thought some of our men were up there, because we had lost some, so I got transportation to the hospital immediately. There I found one of our two lieutenants who were shot down on D-Day. He seemed very happy to be alive. There were other Americans and they all were really thankful for the treatment they had received and happy because the Americans had finally come.

Just before lunch, who should walk into division head-quarters but my squadron commander, Captain Reed (Capt. William R. Reed, Marion, Iowa), who had been shot down on D-Day. What a reunion! He told me his story, about living in a lighthouse for eight days without water, and being a prisoner for six days.

The next morning, Capt. Reed and I started hitch-hiking our way back to an air base. On the road to Sciacca, we drove through very rough mountains. We saw where strategic bridges had been destroyed and found the road full of tank traps. It was a perilous trip.

Further on, toward Licata, we found strategic bridges had been blown up and there also saw tank traps. At one

place, a little railroad town, the place had been completely destroyed by bombs. There were several sidings and two trains, one of gasoline had been hit and burned. The railroad tracks were sticking up in the air like spaghetti.

The trip back to North Africa was uneventful. At our first stop in we met a member of our outfit. He told us that our group had celebrated their 100th day in combat, had flown 100 missions and shot down 100 enemy planes. Believe me we were sure glad to be home - even if it was only Africa. Source: The American Legion Magazine December 1943.

# A MID-AIR COLLISION OF
# TWO B-24s
### by Donald R. Tuohy

It was a contest between my mother and step-father whether we would grow up in Phoenix, AZ or San Diego, CA. On behalf of my brother and me, my mother won that argument. After going to Horace Mann Junior High School and Woodrow Wilson Junior High School in San Diego, I was 15 when the Japanese attacked Pearl Harbor. I was astride my bicycle when my mother shouted to me: "Come in and listen to the radio, we are at war with Japan and Germany." At the time San Diego was a Navy town, and it was rumored that "Japanese cities were eminently flammable, and within a few weeks the U.S. Navy would set them on fire, and the war would be over."

They camouflaged the aircraft factories shortly thereafter. I remember Consolidated Air being under a net and festooned with barrage balloons and, I presume, gun emplacements at pre-set intervals. During 1942, I joined Civilian Defense forces as a "messenger", and I finished my junior year at Herbert Hoover Senior High School.

During my senior year at Hoover, I got a half-time job in the tooling department at Ryan Aircraft Company. I took aeronautics and physics, but I received better grades in Mr. Green's A Capella choir, and Mrs. Grogran's structural art classes. At Ryan we worked on jigs to build outer wing panels for the B-24 bomber, and a hybrid jet and conventional-powered job, the model 828, for the Navy. I graduated in June of 1943, when I was 17 1/2.

After graduation, the part-time job soon developed into a full-time one. I was 17 then, and could have joined the U.S. Navy, but my brother was in the Navy, and I wanted to choose something different. I did take the test to become a naval officer, and I did pass the written examination to attend the Navy V-12 college program. But it was not be to; I was cut out on the quota from San Diego County.

So just one week before I was to turn 18, I enlisted in the U.S. Army Air Corps on Feb. 9, 1944. Three months later, my tooling buddies at Ryan signed my going-away card with "Get a Jap for me," and "Watch out for rear attacks," and with the collected pool of $16.50 for my "widdows and orphants" they wished me: "Fifty cents extra will be given per sub-scriber for the addresses of the children's mothers."

In April of 1944, I was off to Ft. MacArthur, CA. After a tedious train ride to Amarillo, TX, where I took basic train-ing, my next stop was at the aerial gunnery school at Harlington, TX. After aerial gunnery school, I was assigned o Lincoln, NE to pick up an aircrew; and after that phase training, first at Casper, WY, and then at the Army Air Base in Pueblo, CO. All of this took 11 months to accomplish; to train ten men, six enlisted men and four officers, to be a functional aircrew suitable for combat against a wily enemy.

Something out on the tarmack didn't seem to be quite right on that memorable day, March 1, 1945. We were within two weeks of finishing training, and rumors out of headquarters had us heading for Italy to be replacement crews. Staff Sergeant Mark Morris was to be out gunnery instructor that day. We enlisted men had quarreled among ourselves, and I have forgotten, I think it was Staff Sergeant Hoffman, the armorer-gunner, and Corporal Stroud, the sperry ball turret gunner, had to be separated from one another, and our co-pilot, Lt. Frazee, had argued with the pilot that morning, so everything was not right with the world, and our crew, in particular.

The morning passed in a dilatory manner in a series of delays, and the sun was up and brightly shining when at last we got off from the ground. We had barely joined the slowly-circling formation about 20 miles northeast of Pueblo when I, in my tail gunner's position, noticed another B-24 bomber sliding along the right side of us. He was at 2:00 high, and I thought he was flying in formation with us. All of a sudden he started drifting, unconditionally and inexorably, towards us. As he kept coming, I panicked and pushed the intercom button, and I blurted out: "Go down! Go Down! For God's sake, do down!" It was too late. I heard the sickening sound of metal being chewed to smithereens right above my head! And at that moment the other co-pilot looked at me (rather a bit too blase', I thought), as the other B-24 jerkily pulled away. We were at 8,500 ft., and we could feel our B-24 enter a steep dive, and accelerating rapidly, trapping all of us aboard her in an unyielding grip of death. I was trapped in the tail turret, and could not open the doors to the turret, and tried with all my strength to do so. In that moment of the steep dive, I thought to myself: "Well, this is it. You haven't lived, yet you are going to die." All of a sudden the pilots gained control of the errant B-24, by using the trim tabs to pull out from the dive, and the B-24 resumed its nearly normal flight. But the pilots were having trouble. Both top halves of our vertical stabilizers and rudders were missing in that brief moment when the two bombers came together. And now the pilots were losing control. The co-pilot gave the command to bail out. In the interim period, with the pilots in full control, we had opened the camera hatch and the bomb bay doors, and each of us had clipped our parachute onto the harness worn by each of the gunners. How that other gunner beat me out of the ship, I do not know to this day. I was second out of the camera hatch. We did a forward roll, and I saw the B-24 going by, and I knew that I had cleared it.

I was tumbling head over heels, a patch of blue sky, a patch of the earth with its browns and greens, then the blue sky again, kept alternating before me. It was time to pull the ripcord. I gave it a tug. Nothing happened. I gave it another, stronger tug. Nothing happened. A third try, and this time I gave it all of the strength that I had. I had wanted to save the ripcord handle as a souvenir, but the handle went flying off into space, as I watched the chute unfold.

All of a sudden I was suspended in mid-air. After all the noise of a functional heavy bomber during WWII, how quiet it was! I had made a successful jump, and now I had to watch out for telephone wires and poles, and barbed wire fences. We had practiced how to land and with the ground impact we had learned how to do a forward roll. I should have missed that training lesson. I was going too fast, and I hit the ground with my feet first, then my tail, then my back, and finally my head. I had landed in a corn field owned by D.R. Rumsey and Ted Russel of Boone, CO.

Meanwhile, the B-24 had gone out of control and crashed and burned nearby. Everybody had gotten out of it, and the only casualty was our navigator, Lt. Walla. He had a broken leg, as he had landed in a gully and had twisted his leg at an awful angle. The other B-24 landed safely with two bent propellers to show for the collision.

Years later, whenever I would mention that my crew became members of the Caterpillar Club because of a mid-air collision of two B-24's, I got as a response: "yes, but nobody was shooting at you on the way down," (as they did in the European theatre). My crew was later transferred for phase training on the B-32 very heavy bomber out of Ft. Worth, TX. I would up my military career in the Marianas Islands at Tinian, Saipan and Guam, where I was attached to the 21st Bomb Squadron of the 501st Bomb Group of the 20th Air Force.

# MY JUMP STORY
*by Major S. J. Wigley*

It all began when I was scheduled to leave my home base, Biggs AFB, Texas on 11 April 1956, and I proceed to Moses Lake Washington to pick up one of our C-47 aircraft that had been left there. We were to be ferried to Moses Lake in a converted Douglas B-26 bomber.

At our pre-flight briefing I was put in the glass nose of the aircraft and my co-pilot and engineer were placed in the rear compartment. We were to make a refueling stop at McClelland AFB near Sacramento, CA. Our route was via Tucson, Phoenix, Barstow, and the Techachapi Pass into the San Joaquin Valley and northward to Sacramento. Over Bakersfield we received an altitude change to 9,000 ft. We were to have excellent weather all the way. Near Fresno we went into a solid cloud layer and soon were flying in snow and freezing rain.

Near Merced we lost our navigation radios. At this point the pilot called Castle Tower and asked if they had VHF DF equipment, as we had only the old four channel VHF radio for voice communication. Castle tower responded that they had only UHF DF equipment. The pilot did not tell them we were having any kind of difficulty. He continued on toward Sacramento for a few minutes then turned west and after a couple of minutes he then turned south. At this time I knew we were hopelessly lost.

After a few minutes on a southerly heading the pilot began to literally fly in circles. Suddenly we ran into a large hole in the clouds and began a descent down to approximately 4,000 ft. During this descent I found a four land highway and asked him to descend a few hundred feet further so we would see it better and to follow it. I knew if we did so we would find an airport in a matter of minutes. He replied that we were cleared at 9,000 ft. and that was where we belonged. With that statement he proceeded to initiate a climb right back up into the clouds. At no time did he set up any kind of maximum endurance power setting but continued at maximum cruise.

While climbing back to 9,000 ft. I rechecked my parachute harness to be sure it was very tight, took some sectional charts I had and placed them inside my flight suit and zipped it up tight followed by my flight jacket. Finally another half hour of flying in every direction for a few minutes at a time the pilot declared an emergency and a GCI site who had been following our erratic flight answered him immediately and gave us a heading to a nearby airport. Immediately following this transmission the pilot told us we were running out of fuel and stated, "Let's get out of here."

When this bailout order came I face the rear of the aircraft, released the escape hatch in the floor, dropped my legs and lower body into the hold, then pushed myself out against a very strong wind created by an airspeed of over 275 mph.

Since I exited the airplane face up I could see the spinning propellers on either side of me. I was already holding my arms very close to my chest and cleared the propellers without any problem. As I exited the airplane and fell clear the sectional charts I had so carefully stowed inside my flight suit came out the collar and nicked my nick and face in several places. As I fell clear I rolled over to a face down condition grasped the chute harness just below the D-ring with my left hand, place my right hand on the D-ring, looked at it to be sure, then straightened out my right arm. As I felt the ripcord pull the parachute pins and release the pilot chute I decided to look over my left shoulder to watch the parachute deploy. As I did my body rolled and the shroud lines and parachute canopy dragged across my left leg literally burning the skin off from knee to ankle.

I looked at my watch to see what time I bailed out as I knew I would have to testify at an accident investigating board. Following this I grasped the risers and stopped my oscillations. Following this I notice snow flakes at my feet were coming up and hitting me in the face. This caused a good case of vertigo until I realized my sink rate was faster than the snowflakes. Right then I descended into an area of freezing rain. I spend some of my time shaking the parachute canopy and sheets of ice slid off of it. I also broke ice from my face, hands and clothes. Approximately two minutes after I bailed out I heard our aircraft hit and explode. This told me I was still several thousand feet up. I continued to fight the ice on the parachute and myself until I suddenly was below the icing level and descending in rain. At this time I heard what I thought to be a four engine airplane in front of me and to my left. The sound of this airplane became louder and louder and then faded away to my right. I learned later than I was in the middle of an airway and that an airliner had passed at the time I was coming down.

I concentrated on looking downward in the clouds not knowing when I might land or where and trying to be prepared. Then the clouds became darker and all of a sudden I broke out into the clear about 2,000 ft. above a green meadow. The first thing I saw was a white farm house, a barn, a paved highway and telephone lines about half a mile in front of me. I looked around and saw mountains disappear into the clouds all around and just past the farm house. I actually landed in a low area of this meadow where there was a couple of inches of water and lots of mud.

After landing I proceeded toward the farm house, joining my co-pilot who had just landed in front of me. The farmer and

his wife were at home and he knew just what to do. He was a WWII bomber pilot and had heard what he immediately deduced as the airplane crash. After one quick phone call he took us to the Paso Robles airport and turned us over to the FAA controllers there. Following this I called Hamilton Air Rescue, my home base, Caste AFB, and the local police to search for the three remaining occupants of the plane.

A search was immediately undertaken by local authorities for the others. My flight engineer was found in a short time and was hospitalized with a broken shoulder. The pilot of the B-26 was found with a broken left arm, several ribs on his left side and his left leg broken in two places. The B-26 flight engineer was found with the back of his head missing and his parachute not deployed. He and the pilot both went out on the top of the wing and were obviously hit by the tail of the airplane.

# MY LAST BOMBING MISSION OVER NORTHERN YUGOSLAVIA
### by L.G. (Bud) Ballard

It had been a bad day from the beginning. Our S2 briefing information was completely erroneous. We were told there were good bombing conditions in the target area. We found the target to be completely obscured by clouds. We were told there would be a minimal amount of inaccurate flak, we found ground fire was very intense and extremely accurate. They said we could expect very little action from a few Me-109 fighter planes that were stationed in the area, we got swarmed by 40 or 50 Me-109s and the AXIS power's most powerful fighter plane - the FW 190s.

Our plane, number 831, the "Banshee," got hit more than a few times by both fighter plane fire and flak from ground fire. The tail section was practically blown off just in front of me, (I'm flying tail gun). We had numerous holes in our right wing from flak and it was on fire.

We were going down deep in enemy territory. The rest of our crew of ten men had already left the plane. I don't know just when they bailed out because my communications as well as my oxygen system had been destroyed.

I also don't know how long I'd been knocked out, but when I regained consciousness the tail section was a mess. Blood was all over the place and to make matters worse, I knew it had to be mine. I had been hit in the head and face, both legs and my left arm.

I managed to gather up my gear, snap my parachute on and bail out. Knowing I had to get on the ground as soon as possible to avoid capture, I delayed opening my chute as long as I dared to do so. I misjudged my distance from the ground and was well over a thousand feet up when my chute opened.

A Me-109 fighter plane came at me at once and I thought I would be shot hanging in my chute harness, instead he circled me as tight as he could turn. I saluted him, and darned if he didn't salute me back and fly off.

This was my first and only parachute jump and I hit the ground very hard. My head snapped forward, my chin hit my knees and the breath was knocked out of me.

When I was able to breath, I found my mouth full of what I thought to be sand. It was actually small chips from my teeth. I tried to stand up but was unable to do so and crawled and drug my parachute about 50 ft. to the edge of a clump of trees and brush that had a small stream running through it. I hid my chute as best I could with leaves, grass and some fallen branches, then washed my face in the stream. It stung and felt good at the same time. At least I got some of the blood off my face and picked a few pieces of glass and metal out of my face as well.

I used my trench knife to cut a small tree down (about an inch and a half in diameter and six feet long) to use as a prop to get me on my feet. Using both hands on it I was able to move along, not very fast at first, but I got better as I got used to handling the darned thing.

What worried me most was that I had not stopped bleeding from my head, arm, and both legs. But I dared not stop moving. It was 1650 hours when I hit the ground so there was a lot of daylight left and I knew I would be hunted. I opened my escape kit and found my small thumbnail sized compass to be inoperative. The needle was stuck down and wouldn't move.

I knew I wanted to go south and having had a short course in celestial navigation, it didn't take long to chart a course and get started. I didn't get very far until there was a lot of action taking place in the area I had just left. Trucks, motorcycles, and men were swarming all over the hills.

I came to a small creek or river and since the Partisans had destroyed all bridges across streams of any size, I just went right in and waded across. The water was as cold as ice, coming out of the snow covered mountains, that's what it was, ice water. I did this quite often in the next few days and weeks, and it served me well because the dogs they were using to track me with, didn't like to get in the water and that slowed them down a lot. The only time I had a real problem with the water crossings was when I came to the Sava River, one of the largest and certainly the longest rivers in the country. It was wide and in places so deep that I had to do a lot of swimming to get across. Fortunately by this time my wounds had all healed up pretty well and I was a good and strong swimmer.

I finally got with Marshall Tito's Partisans and we were quite busy avoiding the German army, the Micailovich Chetnicks, the Vatican backed Ustache regime, and just plain bandits. Yugoslavia was a very difficult place to live in in those days. My heart went out to those who risked everything to help the downed flyers during the war.

After walking many miles and losing 40 pounds I got picked up by one of our C-47 rescue planes in an open field at midnight. I was returned to the States and got 60 days of medical and psychotherapy treatment, then was assigned to a training command in Armore, CA.

That is where all my military records and citations caught up with me. I received the Silver Star for destroying four enemy fighter planes on a single mission, a Purple Heart, three Bronze Stars and the Good Conduct Medal.

I spent the rest of my military career as an engineer instructor flying B-17 bombers and training crews for overseas duty.

## EMERGENCY JUMP STORY

*by Robert A. Barnum*

During the later part of April 1945, the 57th F.G., then stationed in Grosseto, Italy, was assigned the mission, in conjunction with other fighter bomber squads of the 12 TAC in Italy, to attempt to stop the German army in northern Italy from getting their personnel, heavy guns, allied gear and all forms of transportation from crossing the Po River.

The wheels of the 12th TAC and the 57th F.G. felt that the greatest utilization of these squadrons participating in this mission was to fly all two ship shows (missions) since the turn around time for two P-47s to refuel and rearm was much faster than the turn-around time of four or eight ships. It was also felt by the squadron pilots that since there was very little German opposition during this period to worry about, that two ships whose primary responsibility was low level strafing of M/T and other rolling stock, to include trains, were much more flexible and afforded better utilization of the force, especially since all missions were for "targets of opportunity" on both sides of the Po River.

On this particular day, April 26, 1945, I flew on a two ship show early in the morning with fairly good results (my 132nd mission) since the German army was being pressed by both the 5th U.S. Army and the British 8th Army. As the Germans approached the River, huge road blocks resulted since there were very few bridges across the river that were intact.

During the morning mission there was a build-up of clouds south of the Po and over the mountains as far south as Fiernze (Florence) but the tops were just under 12,000 ft., which presented us no problem, while the area north of the Po, which is a broad plain, was clear with visibility unlimited.

My squadron flew all day with two ship shows and most of the pilots of the squadron had at least two missions that day. About 1630 operations said they needed two pilots for the last flight of the day for the squadron to the valley. I said I would go and my wing man was my good friend and assistant operations officer, Captain Paul M. Hall.

We were off the ground about 1700 and headed north. About 50 miles north of the base we had to climb to 1,100 ft. to get on top of a heavy cloud layer, since it was impossible to go under this deck because of the mountains throughout the Florence and Bologna area. After another 50 miles north, we encountered severe cloud walls and since we were told by our squadron briefing officer that the valley was still clear, we started climbing in an attempt to go over the top. We got to about 1,700 ft. and the last cloud wall looked as though it went to 30,000 ft. and above and after talking it over, we decided to penetrate at 1,700 ft. and hopefully break out in the clear over the valley.

After about 10 minutes on instruments we encountered extremely turbulent weather and were encountering up and down drafts that would take us up 1,200 to 1,500 ft. and almost immediately drop us about the same distance. I estimated by elapse time from T.O. that we were somewhere between Florence and Bologna and in this area is a mountain which we called Mt. Taylor (Mt. Cimore, 7103+) which is about 7,000 ft. of pure rocks.

We continued to encounter extremely heavy turbulence and the next thing I knew I was upside down and my artificial horizon tumbled and the directional gyro was spinning. I attempted to fly needle, ball and air speed after I go right side up, however the turbulence was so great I was unable to keep any semblance of normal flight.

I was upside down, next in a screaming dive up to 500+ mph, then almost stalling out in what appeared to be a loop. This sequence of maneuvers repeated itself several times, but not necessarily in that same order. When I hit 7,500 ft. I decided it was time to get out since I didn't want to be a casualty of Mt. Taylor. Milli-seconds later I unhooked my safety belt, oxygen line and radio connections, jettisoned my canopy and made an effort to exit the right side of the cockpit. About that time the ship rolled and I was pinned half in and half out of the cockpit. I was able to crawl back in and tried a little more needle, ball and air speed but the next thing I knew the ship rolled over on its back and when this happened I stood up on the seat and somehow got turned around and was pinned on the armor plate and was looking at the vertical stabilizer. Following this I was thrown clear (so I thought) of the ship and was floating like a feather on the storm clouds. At this time it was black as night.

I found the D-ring and pulled and shortly thereafter there was a crack of the chute and I found myself upside down with one of my legs tangled up in some of the shroud lines.

After extracting myself from this position, I swung like a pendulum, since apparently I was still in the up and down drafts of the thunderstorm. I attempted to work the shroud lines to stop my swinging but was not very successful. At times I thought the canopy would collapse due to my violent swinging.

Although I couldn't see anything due to the darkness of the clouds and to the heavy rain I knew I was descending since the rain drops were going up and were hitting me in the face while I was looking down trying to find some light and the ground.

In what seemed like an eternity I finally broke out of the clouds and was about 150-200 feet above the ground. Immediately I unbuckled and my chest and leg straps and got ready for the landing. I came down fairly easy in a vineyard in which at first glance I saw the three or four Italians pruning vines.

Apparently none of them saw me come down, probably due to the very low ceiling. I gathered in my chute and sat on the ground between the rows, trying to decide what to do. I thought at first I might try to sneak out of the vineyard since I really didn't know which side of the lines I was on, however the workers appeared too close for that and besides my left leg didn't want to cooperate.

In getting out of my airplane I apparently hit the tail or radio mast since I had a hole in my left thigh about one inch deep, one-half inches wide and about four and one-half inches long. After looking at my cut and my blood soaked pants I decided against running and decided just to sit it out and wait until they found me. After about four or five minutes an old Italian workman finally spotted me and then the whole work crew came over to look.

They helped me out to the edge of the vineyard where there was a small country road and about ten minutes later a small British 8th Army weapons carrier appeared. They loaded me in and told me they were taking me to their battalion aid station, however about a mile down the road we met an American 5th Army weapons carrier and I was transferred from the British to the Americans.

The G.I.'s told me they were taking me to their battalion aid station. After a doctor tried to close the wound with sutures and bandages they put me in a Jeep and we headed south to Florence and arrived there about midnight. I went to the 12th A.F. dispensary and after they looked at my leg, the recommended I stay there for the night.

The next day a B-25 from my group in Grosseto came u and took me home. This flight was my 133rd mission. I staye in the Grosseto field hospital for about a month while my le healed and while there some nurse put a Purple Heart on m pillow with some orders.

Before I left Florence by B-25 I inquired about my frien and wing man Capt. Paul M. Hall and they told me he ha been picked up and brought to Florence. They said he was O except he was slightly shaken up after his parachute ride.

In early 1946, my friend P.M. was assigned to a strateg air command and joined a fighter squadron at Biggs Field, Paso, TX. Later this group was transferred to Chitose A Base, a Japanese air defense base on the island of Hokkaid and after about a year there he and his wing man were caug in heavy icing conditions while making a G.C.A. letdown an landing and both crashed their Republic F-84 fighters an were killed.

P.M. was a good friend and an excellent fighter pilot an he is greatly missed. He is buried in Ft. Stockton, TX where h wife was born and grew up.

Just before I left Italy in 2945 for the SW Pacific th squadron Ad presented me with a small Caterpillar pin o bar of red, white and blue. It is a lapel pin and the retainir button on the back is engraved with the following: R.A Barnum AAF 4/26/45 Pioneer.

# THERE I WAS

*by Colonel E. Dale Boggie*

You've all heard those stories that start out, "There I was a 40,000 ft., straight and level, when all of a sudden..." Well, too used to laugh, but after what happened to me a few year ago, I don't find those situations quite so amusing.

In May 1957, I was a first lieutenant, flying at 40,000 f over New Mexico, on a flight from El Paso, TX to McConne AFB, KS. I was flying an F-86-H, Sabre jet, one of the sweetes airplanes ever built. It was a beautiful, bright, sunny day an so far the flight was strictly in the "routine" category. The things started getting complicated.

A bright red glow from an instrument panel warnin light was the first inkling I had that something was wrong. was labeled "FIRE WARNING, ENGINE COMPARTMENT"

I jerked the throttle back to a minimum power setting like it said in the tech order and racked the aircraft into a stee climbing turn to check for a smoke trail. I remember shadin the little red light with my hand to make sure it wasn't the su playing tricks.

By this time I had turned enough to look back and then knew that this was no trick. Black smoke was boiling out from underneath the fuselage near the trailing edge of the win and extended as far back as I could see. Since the engin compartment was only about three feet aft of my seat, I bega to feel very uncomfortable in that cockpit.

I rolled back level and made another check of the instru ment panel. The red light was still glaring brightly and I knew that my time was running out fast. I remembered briefings o emergency procedures that stressed that the fire warnin system was fool proof and to get out in 10 to 15 seconds.

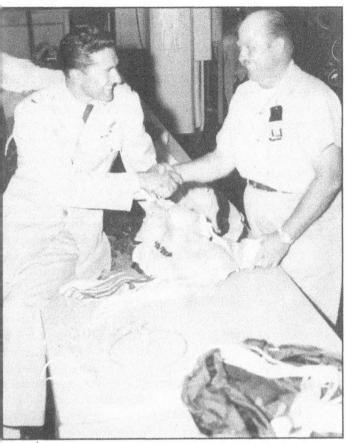
*Dale Boggie with a piece of metal from the F-86-H jet from which he ejected after the jet caught on fire. (Courtesy of Dale Boggie)*

There was no time to make a mayday call. I decided to leave the aircraft before it left me. Emergency procedures came to me automatically. Down came the helmet visor, out came the little green "apple" on the bailout bottle. Up came the arm rests and the canopy exploded off. The plane was still going about .85 mach (or over 550 mph) and the wind blast slammed me up against the seat belt and shoulder straps. I squeezed the ejection seat triggers, felt a terrific jab, heard a whooshing explosion and saw the instrument panel, with the little red light still glowing furiously, fade down and away. That was the last time I saw the aircraft, but I'll never forget it.

The tumbling I encountered was very severe and I felt exactly like a rag doll. Once I saw my legs silhouetted against the sun, but they seemed to be twisted crazily out of shape and didn't look much like legs at all. I reached down to check for the seat belt, but it had already separated and the seat had fallen away.

I finally got my arms and legs collected and stuck out an elbow which helped to stop the tumbling. I was dropping straight down, turning slowly, sometimes face down and sometimes on my side or back.

I didn't particularly notice the cold, although it was in the neighborhood of -55 F at that altitude. There was no sensation of falling like you might get from jumping off a wall or starting down an elevator. I just seemed suspended in space, unable to see clearly and with a terrific roaring in my eyes due to the wind blast.

I grabbed hold of the rip cord, looking down to be sure I had a good grip on it, but forced myself to wait for the automatic device to open the chute. It never occurred to me that it might not open at all. I knew that the chute was set to open automatically two seconds after passing 14,000 ft. That meant I had a free fall of more than 26,000 ft, or about five miles!

Later I wondered exactly how much farther above 40,000 ft. I did go, considering the climbing turn to check for a smoke trail and the blast from the ejection seat that kicked me up and away from the airplane. One thing for sure, when you pull those seat triggers, OUT you go! It had a charge about equal to a 40mm cannon shell. I was pretty touchy about sitting down for a week for so for this and other reasons.

As to what I thought about on the way down, I certainly didn't see my life passing before me or any such thing. I was more concerned about rechecking my mask and jerking on the bailout bottle lanyard again for good measure. I was glad I had developed the habit of connecting the bailout bottle and fastening the chin strap on my helmet on every flight, no matter how high I planned to go. It sure paid off.

I couldn't see the ground clearly and it was hard to judge over that bleak desert anyway. I knew there were 10,000 foot mountains in the area and rechecked my grip on the ripcord. One of the longest waits I've ever experienced ended when the chute blossomed out automatically. Now I know what they mean when they say, "Watch that first step, its a long one." In this case, it was five miles!

That big canopy over my head was a beautiful sight to behold. Everything got very quiet and I raised the helmet visor and unhooked my oxygen mask to look around. It began to look like a nice day again. The ground still seemed a long way down and I thought it would be a long time before I got there. I looked around the airplane but there wasn't a trace. I didn't see any recognizable landmarks either, except the mountains I was worried about. They were well off to the east. I did see a water hole not too far away and decided to head for it when I landed.

Suddenly I could see how fast I was dropping towards the ground. In addition, a strong wind was carrying me cross-county like an express train. I managed to turn myself so I was in a quartering position, and then I hit. It was a tooth rattler! Later I learned the ground elevation was still about 7,000 ft. This, coupled with the way the wind was pushing me, resulted in one of the worst jolts I've ever taken.

I tumbled and rolled, just like the parachute landing falls we had practiced. Other than driving my heels up through my buttocks and taking a nasty crack on the head (which cracked my helmet), I figured I had done pretty well. Then things started getting out of hand again.

I had opened up the quick release on the parachute harness on the way down. However, before I could get the second half of it unlocked after hitting the ground, the chute started dragging me. The wind speed was about 25 mph and off I went, face down, like a human surfboard.

The ground was relatively flat, but rocky, and covered with cholla cactus plants about four inches high. Every time I got my hand on the quick release on one of the two parachute risers. The other riser was permanently fastened to the parachute harness. Theoretically, this meant you would always have your chute tied to you for use in a survival situation. It may have been a good theory, but not worth much when all you were trying to do was collapse the chute to keep from being drug. The side with the quick release was the side I couldn't seem to get to. (Chutes were later modified back to two quick releases.)

Finally I just started hauling in shroud lines. A couple of times I had the chute almost collapsed and got to my knees. Then a big gust of wind would snap the chute out, and off I would go again. I was glad there were no barbed wire fences in the area, but that cactus was sure giving me fits.

I was able to pull in enough of the shroud lines to gather the chute in and sit on it. I looked like a porcupine with all the cactus sticking in me. The spines were those beastly little ones, with barbs like fishhooks on the end.

I took off my helmet and took stock of the situation. No bones were broken, but I was sure bruised and stinging from the cactus barbs. Every solitary thing I'd had in my pockets had been ripped out- wallet, car keys, pencils, maps and charts were all gone, including my knife which I really needed.

I then became aware of the sun beating down and knew I'd better get to that water hole. Survival movies I had seen, warned of the danger of shock setting in unless you had a definite plan to follow to keep you occupied. I sure didn't want to end up in a daze, wandering around in the desert.

After more than two hours of walking, I began to wonder if I was going around in circles. I had picked out a mountain peak as a land mark and kept going in a straight line. The mirages of lakes I kept seeing off to the sides didn't help any, but I kept to the course I had set. It had seemed only a short distance, sitting up in that parachute looking through the clear high desert air. It turned out to be quite a walk.

The water was seeping out of a hillside and running down into a small reservoir. I drank more than any self-respecting camel would thing of. Next I spread the chute out for a signal and weighted it down with rocks. Then I sat down and tried to relax for a while and think of what to do next. I needed a knife in a bad way to get the cactus out. Trying to pull the barbs out with my fingers didn't work very well because most of the barbs broke off and stayed under the skin.

I wasn't sure of which way to go in case I had to walk out to the nearest civilization and figured my best bet was to stay near the water hole. After about three hours, it seemed doubtful that I would be found that day. I started clearing a spot and looking around for material to make a shelter of some sort with the parachute. Then from out of nowhere I heard a jolting, rattling noise which sounded exactly like an old truck.

I hobbled up on top of a slight rise and sure enough, an old tan, Ford pickup truck was kicking up a cloud of dust no too far away. I yelled and waved my arms and the truck changed course and drove up to me. An elderly Mexican was driving and had apparently been up in the nearby foothills looking for stray cattle or something.

He spoke no English but I managed to remember enough Spanish to let him know I needed a telephone. He agreed to take me to the nearest town, which seemed to be 15 or 20 miles away. He looked a little wary when I asked him to wait while I piled my chute and helmet in back of the pickup. I knew I didn't look too good after being drug through the

cactus, so I tried to smile a lot and act friendly. He also had a old 30-30 rifle which he placed near his left leg so I didn make any sudden moves on the trip to town.

"Town" turned out to be Las Lunas, New Mexico, and took me to the courthouse. I thanked him as best I coul dumped my gear on the lawn at the base of a cottonwood tre and walked up the long tree-lined walk and into the buildin This caused quite a stir among all the ladies working in the b room behind the counter. Everything stopped and they ju stared at me. I looked down the hall and saw a sign that sa sheriff's office and ducked in there.

As I stepped through the door, the police radio sitting top of an old roll-top desk was crackling out a messag asking if anyone had any information on an overdue ar presumed crashed aircraft. I recognized the description as m F-86-H and tapped the young Mexican operator on the shou der. He turned around with a great look of surprise on h face. I told him to report that I was all right.

He got very excited and started interrupting all the oth stations on the net. He made it sound like I had brought th plane in for a crash landing right on main street. I tried correct him, telling him I didn't know where the plane wa that it was on fire when I left it and had probably blown u He immediately started interrupting other transmission again. I saw a telephone sitting in another office so I went there and asked the operator to place a collect call to the ba operations officer at Kirtland AFB, New Mexico.

A sergeant answered the call and the operator asked fo the base operations officer. The sergeant said he couldn come to the phone because they had an emergency in progre involving a lost aircraft and to call back later. I interrupted t say, "I am the emergency." There was a moment of silenc and then "Standby sir, I'm sure the major will talk to you.

The base ops officer was very glad to hear from me an wanted to know where I was. When I replied, "Las Lunas", h asked where that was. I looked at a map hanging on the wa and estimated it was 40 to 50 miles south of Albuquerque. H said they had a helicopter out searching, but by the time could get back, refuel and come down for me, he could hav a staff car there. I said that was fine with me and that I woul be out in front of the courthouse.

I retired to the front lawn where quite a crowd ha gathered around my chute and helmet. They were politel curious about now it felt to bail out of an airplane almost eigh miles up. Before long the staff car arrived and I was asleep the back seat before the driver had it in high gear for the tri back. When we arrived at Kirtland, the flight surgeon checke me over and found me fit. I borrowed a pair of tweezers an went to work on the cactus barbs in earnest.

Then it was down to base operations where there was room full of officials with all sorts of questions. I answere them all as best I could. As the day began to draw to a close I realized I was still in some kind of predicament. I was at strange base with no identification, no money, no shavin gear and no clothes, except the ripped up flying suit and slip on Wellington flying boots I was wearing. The base op officer adopted me like a long lost brother and took me hom with him.

The next day my home base sent a T-33 to pick me up and I was allowed to go back to Kelly AFB, TX, with the provisio that I return in two weeks to meet an accident investigation board. During the interim I called Kirtland a few times, asking if they had found anything of the airplane yet. The answer was no.

When the two weeks was up, I flew back to Kirtland to meet the board. When I arrived I was met by the base ops officer, who beckoned me into the flying safety office. There was a cardboard box full of bits and pieces of aluminum, turbine buckets, etc. One mangled up piece of aluminum about a foot square looked like an inspection panel door or part of a tip tank. It had a small stencil on it containing the tail number of my aircraft! It had obviously been involved in a violent explosion.

I then reported in to the accident board and formally testified as to the events which occurred. When the flying safety officer introduced the collection of metal bits and pieces recovered from the desert, my testimony was fully corroborated. Incidentally, the pieces of metal were found by a young boy who was out hunting or something. His grandfather saw the collection and luckily for me, decided to report it to the local police who passed it on to the proper authorities. It arrived at Kirtland just the day before the board met.

I still have that piece of metal, with the tail number stenciled on it. It hangs on a plaque displayed in my den,

along with a scale model of the F-86-H. It has served as a reminder of the importance of training and of developing and constantly practicing safe habits. It served me well through 28 years with the Air Force, flying in peacetime and in combat, and continues in civilian flying. I am thankful to all the people right back to the primary flight school days who contributed on one way or another to my training. Training that gave me the knowledge and confidence to survive a bad situation, when "There I was at 40,000 ft., straight and level, when all of a sudden...!

# HOW I QUALIFIED TO BE A CATERPILLAR
*by Louise Bowden Brown*

Just before World War II I learned to fly in a small field in South Jersey. Then came the war and women were called into the service - Jackie Cochran accepted me for the ferry command. I trained for six months in Sweetwater, TX and was then assigned to the Third Ferrying Group in Romulus, MI.

At first we delivered small planes and then after receiving a white card from instrument school in St. Joseph, MO, I was sent to fighter school in Brownsville, TX. There we had ground school half days and flight line the other half day. We

*Louise Bowden kneels on the plane she ferried in 1944 for the Army Air Force at Chanute Field as she explains her job to a WAC officer stationed there. (Courtesy of Louise Bowden Brown)*

# Young Aviatrix Bails Out of Burning Plane

Louise Bowden, of Elmer, young aviatrix, had an exciting experience while with the WASPS when she parachuted to safety from a burning Army fighter plane near Greensboro, N.C. It was the pilot's first forced jump in more than three years of flying.

According to an Army press relations release, witnesses reported that the plane was in difficulty for a time prior to the crash. Finally Miss Bowden notified the Greensboro-High Point airport that she was in trouble and was bailing out. It was reported that the ship was afire before the pilot bailed out but this was not verified by the Army.

The plane crashed in a plowed field. The single motor was buried in the ground and a minor explosion followed, turning the craft into a roaring inferno within a few minutes. The pilot, who bailed out at about 1500 feet and landed about 20 yards from the burning ship, was uninjured.

Miss Bowden had qualified for a civilian pilot's license before going to Sweetwater, Texas, to fly with the WASPS. She received her training at Buck's Airfield, at Woodruff, with everyone around the field glad to help the little red-headed girl who was so eager to learn to fly. She was a technician

**LOUISE BOWDEN**

at Bridgeton Hospital before making flying her career. The young woman pilot has flown every type of Army plane including the speedy fighters which require great skill to handle.

flew the AT-6 from the back seat with an instructor in the front seat. Engineers from the Pratt & Whitney, Packard, and Alison engine manufacturers lectured us on the operation of these engines with a lot of do's and don'ts. We had excellent training.

Then we were checked out and flew five single engine fighters for two weeks. These were the P-47, P-51, P-40, P-39, and P-63.

Back at Romulus we flew these from factories to training fields or to ports of embarkation for shipment overseas.

I had picked up a Mustang in Texas and remained over night (RONed) in Greensboro, NC. The next morning a group of us took off for Newark, NJ and had leveled off about 5,000 ft. After cruising a short while the engine sputtered. Automatically I was doing a 180 back toward the field and asking the Lord to help me. I called the field to say I was en route back and in trouble. Several more times it sputtered, ran smoothly and then quit. When I released the hatch, the head set went with it. With difficulty I got up and fell over the right side. I remember seeing blue sky beyond my feet and a twin engine Navy plane. I felt that they saw me. When the chute opened the burning plane stared at me. I steered away from it and landed about 30 yards from the flames in a harvested corn field and sank in mud to my shoe tops. Above me was the circling Navy plane and I waved to say I'm OK - later I learned that they called the field I had departed.

Then I saw an Army lieutenant running towards me. He and his wife were looking at the Navy plane and saw me floating down. They drove me to a nearby Army hospital, where I was treated for delayed shock and a scratched ankle. I was aware of being surrounded by people who wanted to

help me - in contrast to our fighting men shot down over enemy territory and abused if captured or injured and alone.

Because I experienced the Lord's gracious love and forgiveness I was led to a conversion and later to serve in the mission fields of India and Nepal.

## POINT OF NO RETURN
### by Peter Cotellese

In 1956-57, the 3rd Bomb Wing at Johnson Air Base, Japan converted from the B-26 to the B-57 aircraft. Lt. Gayle Johnson, navigator, and I formed one of the crews returning to McClellan AFB, CA in April 1957 to ferry one of the last three B-57s destined for Japan.

On April 29, after a successful acceptance flight of B-57 53-3916, three B-57s, with Lt. Col. I.H. (Itch) Young as flight leader, departed McClellan AFB on the first leg to Hickam AFB, Hawaii. Midway into the flight, over Ocean Station November (a weather station manned by a Coast Guard ship), strong headwinds forced us to turn back. I landed back at McClellan AFB five hours and 45 minutes after take-off with 1800 lbs. of fuel remaining. It was a perfect fuel consumption curve.

On May 12, Mothers Day), we tried again. The leg was planned for a 38 knot headwind component (40 knots was maximum) and five hours and 33 minutes time en route at maximum cruise altitude. I was the last to burn out fuel in the wing tip and ferry tanks, at a point 150 miles east of Ocean Station November. A short time later my fuel consumption increased.

Suddenly my airplane had consumed over 1,500 lbs. more fuel than the other aircraft. Hoping the fuel gauge had malfunctioned, I continued in the flight. Minutes later the low level fuel warning light came on indicating I had approximately 5,000 lbs. of fuel remaining; not enough to continue to Hawaii or return to California. We turned back to Ocean Station November to bail out.

The station was manned by the U.S. Coast Guard cutter Wachusetts, under the command of Commander Gerald T. Applegate. Our plan, worked out with ship personnel, was simple. With the ship heading into the wind, we would fly over it at 3,000 ft. and 160 knots. One minute past the ship, we would eject and parachute to the ocean where the ship would pick us up. Simple? Not quite.

When it was obvious we had to bail out, the navigator and I reviewed emergency and ejection procedures. I instructed Lt. Johnson that, on my signal, he would blow the canopy and immediately eject. I would follow. Simple? Not quite.

During the descent to the ship, I thought it would be smart to blow the canopy early, clear the cockpit of debris and retrim the aircraft for straight and level flight. The idea made so much sense, I bought it; but I did not tell the navigator. On our final pass, at a distance of eight miles, we lowered our visors and I asked the navigator to blow the canopy. Pow went the canopy and pow went Gayle's seat. He ejected eight miles from the ship.

I panicked. I informed the ship and turned back to eject near the navigator. I was instructed to return to the ship and eject as we had planned. I did. My seat ejected, it separated

The B-57 and Cotellesse's parachute shortly after he ejected. (Air Force photo, Hickam Air Force Base, courtesy of Peter Cotellesse)

Cotellesse's B-57 crashes into the sea as seen from the ship Wachusetts. (Air Force photo, Hickam Air Force base, courtesy of Peter Cotellesse)

A rescue crew from the Wachusetts moves quickly to pick up Cotellesse. (Air Force photo, Hickam Air Force Base, courtesy of Peter Cotellesse)

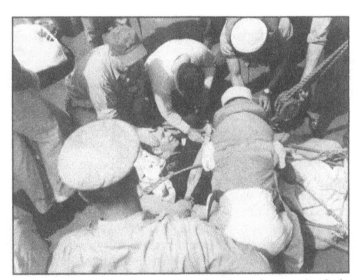

Cotellesse (white shirt) watches as a crewman from the Wachusetts assists Gayle Johnson, B-57 navigator. (Air Force photo, Hickam Air Force Base, courtesy of Peter Cotellesse)

Peter Cotellesse pictured with his B-57 while assigned to the 3rd Bomb Wing, Johnson AB, Japan, about 1955 to June 1958.

from me, the chute opened and I was in the water nine minutes. Thirty minutes later we spotted the navigator's flare at a distance just off the bow. He was in the water 40 minutes.

Gayle Johnson, now retired in Wilmington, NC, and I have remained good friends and almost every year we talk to remember a once in a lifetime experience.

## A FIRING SQUAD AND A GHURKA
*by Walter Crowell*

I was second out of the stricken bomber, right behind the nose gunner. I dove headfirst through the open bomb bay into the rushing slipstream and was whipped back past the right tail of the dying plane. My chute burst open with a tooth-jarring jerk. With my one good eye I watched eight chutes blossom in the air behind the falling plane and felt relieved that all of us managed to escape. Still numbed from the violent flak explosion that savaged the nose of the plane, I tried not to think about being permanently blinded by the shards of plexiglass from the B-24's windshield hurled into my left eye.

Swinging gently beneath a canopy of white silk, I was mesmerized by the abrupt silence and the peace and tranquillity of the moment. This brief interlude was short lived.

Sudden intense pain stabbed through my injured eye. Something sticky and gooey dripped into my flight jacket. It was thick, dark red and icy. I took off my glove and reached towards the pain. My left eye socket was filled with icy slush. It was frozen blood. Warmer temperatures at lower altitudes thawed the frozen blood causing the wound to bleed freely and the pain to intensify.

Escape was foremost in my mind as I drifted slowly down toward the checkerboard farmland below. Moments after I hit the ground and struggled to free myself from the chute, I was surrounded by a group of irregulars, the People's Police, called Volksturm, armed with pitchforks, fowling pieces, shovels and similar weapons. They worked me over with kicks and beatings and screamed insults and obscenities which I could barely understand. Though I was cut up, bruised and bleeding, they stopped short of making me a cripple. Then three men in green uniforms arrived. These members of the constabulary drove away my tormentors and took me into custody.

My captors gave no sign of understanding English and cuffed me into silence when I tried to speak. They pushed me into a tiny, beetle-like Kubelwagon and we drove a short distance to their headquarters, a small police station. Inside, two guards grabbed me by the shoulders, lifted me clear of the floor and threw me through an open door into a darkened room where I fell on the cold concrete floor. The heavy metal door slammed behind me and I heard a key turn in the lock. It was deathly still.

I was alone in a surprisingly large holding cell, about 12 by 14 feet. The only light came from a small window at ceiling height in one corner of the room. In the dim light I could make out four wooden pallets used for beds along with wall. I sat on one of the hard beds to ponder my fate.

Within a half an hour of my capture, the Germans had rounded up the nose gunner, the engineer, the tail gunner and the two waist gunner and tossed them into the cell with me. As first lieutenant, I was senior officer. I reaffirmed that everyone had parachuted from the plane, and took note that I had sustained the only injury.

Our hushed conversation plotting escape was interrupted when the door of our cell was flung open. Outside there was loud shouting and the guard at the door, armed with a rifle, motioned with the muzzle of his gun for us to come out. "Raus! Raus!", he shouted. Nine other green uniformed men waited for us outside. Jabbing us with their rifles, kicking us and beating us about our shoulders and rumps, they forced us through an open doorway into a walled-in courtyard. We were pushed over a side wall and ordered by hand signs to face our captors.

The squad leader shouted and gestured to us to about face. When I turned around and faced the wall, I saw directly in front of me that the whitewashed wall was pitted with dozens of bullet holes exactly at the level of my chest. Thoughts of dying in a thousand different ways, always in the air, were a part of every combat flier, but death by a firing squad had never crossed my mind.

The squad leader ordered his men to attention.

"Achtung!", he shouted.

The sound of ten rifle bolts slamming home shattered the stillness in the courtyard. In the deadly silence which followed, I could feel the center of my back framed in the sight of ten Mauser rifles aimed directly at me.

"Feuer!"

Ten firing pins clicked metallically inside empty rifle chambers but the sound was almost as loud as live rounds. One of the crewmen had dropped to the ground, legs drawn up in the fetal position, rocking and moaning. I was too frightened that I'd be shot to go to his aid.

I turned to face my tormentors, but they screamed at me to face the wall. I heard them re-charge their weapons. This time I was sure they rammed in live rounds when they pushed their bolts home.

"Achtung!"

"Feuer!"

I don't know how long it took the Germans to tire of their fiendish game of mock execution. Finally they kicked and beat us with rifle butts back to our cell. Inside the dark room I groped my way to one of the wooden pallets and dropped down on the hard boards feeling as dead as if my body had been riddled with bullets. I began to shake and tremble uncontrollably.

Footsteps sounded outside the door and the head guard opened the door and beckoned me to come out. There was change in the guard's demeanor. He acted less hostile, even respectful. He brought me into the main reception area. The setting sun sent a shaft of bright orange light streaming through the open door leading to the outside world and freedom. In the doorway, their faces and bodies silhouetted against the brilliant sun, stood two men. As I drew near, I saw they wore blue uniforms. One was a Feldwebel, the other a lieutenant. When I approached, the Feldwebel drew himself to attention and the Lieutenant snapped out, "For you the war is over!"

For me, the war as I had known it, raining death and destruction from the skies onto the enemy below, was indeed over. I was now in the hands of the Luftwaffe, raw material for the efficient prison machine they had built to process captives through a pipeline beginning at point of capture and ending at one of several Stalag Lufts located mostly around the eastern parameters of the Reich.

For the first time since bailing out of the bomber, I felt I had cheated death. The two Luftwaffe personnel sent to pick me up were disciplined, professional soldiers. They made it clear to the armed men in green uniforms that their job was to get me to prison camp in one piece, not killed along the way. They brought me to a doctor who treated my injured eye.

Although I didn't know it at the time, I was to embark on a two day train ride to the Luftwaffe Interrogation Center at Frankfort am Main. Two enlisted men were assigned to escort me. Each was armed with a pistol. I never learned their names nor could I determine their rank.

Fear and anxiety remained my constant companions, but the train ride to Frankfort in a third class compartment aroused my interest. It was an opportunity to view the enemy first hand and be a witness to the devastation and destruction caused by Allied bombings. I didn't travel through a city that was unscarred. It was a wonder I thought that Germans were able to keep the trains running.

At the start of our journey we were in farmland. Old men

and women and children boarded the train toting baskets full of produce and handcrafted goods. Some baskets served as cages for geese. There were openings in the tops of these cages and the birds' long necks would stick up in the air bringing their bright orange bills even with their owners' noses. One elderly man, white bearded and elf-like as St. Nick, held a cage of squealing piglets on his lap.

Nearing more urban area the mix of passengers changed. Business and trades people filled the compartment. Luggage, brief cases, tool kits and reading material were carried aboard by these city folk. In the country people looked at me in wonderment. Near cities I was viewed with anger, hostility and resentment.

On this rail journey behind enemy lines, I experienced three "firsts": my first meal, my first air raid and my first Gurkha.

It was just past noon when the train slowed and edged into a station. Thirty hours had passed since I'd eaten and I was famished. My guards motioned me to get off the train. The size of the station told me we were at a major rail center. We walked to the far end of the large yellow brick structure and entered a room that appeared to be a large cafeteria. Everyone inside the hot, steamy room was in uniform. We were in an armed forces canteen.

My escort pushed through the crowded room and found three vacant seats at one of the large tables. One of the guards left to fetch food. He returned with two large soup bowls filled to the brim with a rich beef stew and two large chunks of black bread. He set one bowl in front of his comrade and took the other one for himself. The two men quickly finished their meals.

I did my best to ignore my guards who obviously were enjoying their meal, but when the older of the two looked my way and pointed to his empty bowl I eagerly nodded my head. Soon there was a steaming bowl of stew in front of me. Full of meat, potatoes, and carrots, with rich brown gravy, it would be good stew anywhere in the world, whether in war or peace.

With a full stomach my spirits rose and I didn't feel quite so hopeless. Exhaustion caught up with me and when we boarded another train I fell into a deep sleep. I don't know how long I slept nor the exact time of day when the train pulled into the main terminal at Frankfort. From the lengthening shadows it must have been late afternoon. We stepped off the train just as the air raid sirens began their banshee wails. The only sirens I had every heard came from the volunteer fire department back home. These sirens were different. Their deafening sound seemed to emerge from the walls right inside the train terminal. I felt confused and frightened by the piercing noise and wondered what it would be like when the bombs rained down from the sky.

Everyone on the crowded platform quickened their pace and hurried towards the terminal. It was an orderly procession. No one ran. There was no pushing or shoving. Jostled along as we were by the crowd, I felt naked and bare from the hostile glances thrown my way by civilians and the occasional outbursts of profanity. I felt the danger and was relieved the way my guards hurried me along and shouldered their way through the crowd.

We made our way into the terminal and down a steep flight of stairs several levels underground and into a dimly lit corridor. At the end of the corridor was a locked door. The door was opened and I was pushed into the room beyond. The only light came from high up on the ceiling where a single bulb barely gleamed through a thick covering of soot and grease. I was in a big room; it must have been 20 by 40 feet. Every square inch of space on the floor was covered with human bodies. The stench of the unwashed bodies, excrement and general filth washed over me. I had to swallow hard to keep from vomiting.

Through the gloom I saw a raised arm beckoning to me. The arm belonged to a wild-looking, bushy-bearded man with a filthy turban covering his unruly hair. He roared and bellowed commands I could not understand, but a way was opened for me through the sea of bodies to where he was sitting at the rear of the room with his back against the wall. He motioned for me to sit beside him and spoke to me in pidgin English. Combining his broken English with sign language we were able to carry on a conversation. My benefactor recognized me immediately as an American and considered it an honor to have me seated by his side. I learned he was a Gurkha. He told me he was captured in the desert y Rommel's forces in 1941.

A series of violent explosions shook the room. Even deep in the bowels of the terminal building we were rocked again and again by the forces erupting outside. I looked around wildly like a frightened caged animal and asked how close the bombs were falling. I felt I was losing control and became claustrophobic in the densely packed room. No one else appeared frightened and those people asleep on the floor never stirred. I repeated my question about the bombs. The Gurkha laughed and told me all the noise came from flak guns. The bombs hadn't started to fall.

After a time the guns fell silent. No bombs fell. Just before the all clear sounded, the door opened and two Germans entered struggling with a heavy pot of steaming food. It looked to me like soup. Out of nowhere everyone in the room, except for me, held out a cup or other container. The Germans hurriedly passed through the crowd pouring out one ladle of soup for each person.

Hunger had wasted my defenses. I coveted the chipped porcelain cup my Gurkha friend held out and I was fixated with the thin, watery gruel pouring from the guard's ladle. I tried to tear my eyes from the steaming cup being raised to the Gurha's lips, and I saw him hesitate for an instant. He turned and peered directly at me and I averted my eyes abruptly feeling guilty like I had been caught staring at a voluptuous woman. The Gurkha asked me when I had last eaten. I held up two fingers for two days. Without a word he thrust the brimming cup into my outstretched hands.

## A WEEKEND RIDE
### by Mae Belle Hardin

My jump story is quite tame compared to most others. I was a control tower operator at Perrin Field, Texas at the time, loved flying and went along "for the ride" whenever I heard of an empty seat, going most anywhere.

The weekend of my misfortune, I caught a ride with a pilot that I never met, who was headed for Georgia to visit his family, and he dropped me off in Birmingham to spend Mother's Day with my family.

On the way back to Texas we ran into bad weather, the radio went out, and we circled for about two hours trying to find a hole in the overcast and land. On the way the fuel warning light flashing, we bailed out at about 7,000 ft. over a solid overcast, about 2:30 A.M. with no idea where we where.

The pilot didn't know if I would jump, and frankly I was so frightened I didn't know if I could either. I owe my life to the crew chief of the plane who helped me tighten the parachute harness before I left Perrin.

I weighed barely 100 pounds in those days and even then, as I drifted down in total blackness, I felt that the harness wasn't "right". Both shoulder straps were on one shoulder - in tumbling down or when the chute opened, I had come through part of the harness. I remember thinking I should put the thing on right - then decided I had best just hang on and leave it alone. Shows you I wasn't thinking very clearly.

Due to the solid overcast and the fact that we jumped in a rural area near a very small town in Oklahoma, I couldn't see the ground until I hit it. I swung into a high power line pole on the descent, scraping one leg, and landed in a ditch paralleling a country road. Fortunately the only damage was a cracked rib and skinned leg.

I walked to a nearby farmhouse and the family took me to a neighbor with a truck who drove me to Hugo, OK.

The pilot fell in a tree and skinned himself up, but neither of us were really hurt. A week later I had an opportunity to fly back to Birmingham so that my parents could see that I really was all in one piece.

I managed to hold on to my ripcord - I had such a grip on it that I didn't realize I still had it until I hit the ground.

Our oldest son reminded me recently that his very favorite thing to take to school when he was in elementary school was that ripcord, and he would have an opportunity to tell his classmates on "show and tell" about his mother's parachute jump. After several years of this, they would groan "Oh no, not again" when they would see Mike stand up with that ripcord.

## FIELD TRAINING BAILOUT
*by Col. Roger H. Hebner*

The 115th Observation Squadron, 40 Division Aviation, was in annual field training encampment, July 12, 1932, at Camp San Luis Obispo. I was a private first class, specialist second class, and a crew chief nearing the end of my first enlistment. First Lt. Charles W. Haas was the squadron engineering officer. The 0-17 airplane had just been returned from a major overhaul at Rockwell Air Depot at San Diego. On the morn-

*Wreckage of Consolidated O-17 at S.L.O. A/C 28-372. 2nd Lt. Charles Haas and Pfc. Roger Hebner bailed out when right upper wing crumpled during aerobatics. 12 July 1932. (Official photo, 146 TAW, courtesy of Roger Hebner)*

ng of July 27th, Lt. Haas was to make a series of test flights. was aboard the second or third of those flights.

We took off and flew towards Morro Bay under a ceiling of about 2,500 ft., coastal stratus. We flew up into the stratus and then started a power dive and just after coming into the clear the plane whipped about violently. My thought at the moment was, "He's really wringing it out!"

Almost simultaneously something slapped my face (undoubtedly some fabric from the wing panel), and I ducked down in the cockpit. As I looked up I saw the heels of a pair of boots going upward out of my field of vision. I flipped my safety belt and started to put my feet on the seat to push out, but settled back long enough to physically locate the rip cord handle with my right hand to assure myself that it was actually there. I then put my feet on the seat and pushed out. Clearing the airplane was no problem as I saw it spiralling rapidly away below my G.I. boots.

I pulled the ripcord and as the chute opened and brought me to what seemed to be a standstill, I had the feeling of "How great it is!"

I looked down for Lt. Haas' chute, finally finding it well above me. I was hearing the scream of the plane for about ten seconds after the chute opened when it ended in a loud crunch. Then silence prevailed.

Watching for my landing I saw one large rock formation which concerned me, but the coastal breeze kept me away from that but blew me into the upslope of the hill for a landing that bounced me on the seat of my pants and flipped me over to drag me up the hill. I spilled the chute, rolled it up and with the ripcord still on my wrist, walked toward the crest in the direction where I'd last seen Lt. Haas in the air. He met me at the crest, having landed downslope. Both of us were extremely relieved that neither was injured. We then walked down toward the wreckage where the plane had mashed itself in a natural ditch.

One other recollection that I have, during my descent I saw the one wing panel fluttering down like a leaf, trailing confetti. That's when I really knew what happened.

## VALENTINE'S DAY BAILOUT

*by J.W. Holland*

We had returned to our (H-74) field at Camberia, France. The gear had unlocked and was 27% down, in the landing pattern, when asked to proceed to the Channel to dispose of two 500 pound bombs remaining in the bomb bay. The bombs were hand jettisoned on some farm land, when the pilot and navigator realized there was not enough fuel to make the Channel trip and return.

Fuel was used to try to completely lower the gear, when it was found the hand cable was connected or had been cut into by shrapnel, thereby decreasing our remaining air time. A frozen hydraulic pump on air pocket in the line prevented the full pressuring.

The navigator's chute did not flare. (Roman candeled.) He was dead when we got to him. He had used my pack. I had used his. I'm looking for a pack and he used the nearest one available.

I was the last man to leave the plane. I dropped through the nose wheel well.

The pilot and co-pilot went in with the plane. The pilot was

*Edward J. Kenavan and Jackson W. Holland, 1944 in Ireland. Both bailed out of their planes on Feb. 14, 1945. (Courtesy of Jackson Holland)*

killed. The co-pilot was burned some time after hospitalization returned to pilot a crew before May 9, 1945.

A Frenchman carried me to a village in a manure cart pulled by an ox and a horse.

B-26 number was 217 of the 585th B.S. of the 394th Bomb Group. In landing it hit a four foot high incline and came into just aft of the rear edge of the wings.

## EVENTS LEADING TO BECOMING
## A CATERPILLAR

*by Benjamin E. Hopkins*

On May 2, 1945 as instructor-pilot, the crew I was training and I had a close call.

While flying at 25,000 ft., the left scanner (gunner in time of war) called me and said number one engine was on fire and the flame was coming back to the horizontal stabilizer.

I had the plane pressurized to a cabin altitude of 8,500 ft. The emergency cabin release would not work, so I told the flight engineer to start depressurizing manually as I nosed the plane downward. At 20,000 ft. the pressure was equalized, so I ordered the crew to abandon the plane.

As instructor airplane commander, I was the last one to leave the plane. I jumped at 15,000 ft. The parachute failed to open when I pulled the ripcord.

I pulled the parachute, pilot chute and shroud lines out by hand and became tangled in shroud lines, etc.

After falling about two and a half, I captured a little air; three small canopies opened.

The terminal velocity of a human body is 120 miles per hour. I hit the ground hard, equivalent to a 50 ft. free jump, so I was told by parachute experts.

I suffered a broken neck, broken left ankle and internal injuries. After spending about eight months in three different hospitals, I was discharged and sent to San Antonio, Texas for reassignment.

The V.A. pays me 50% disability.

## The Story Of My Bailout

*by Albert W. James*

My flying school classmate, Lt. Jack Alston, had been visiting family in the Ogden area, and was in need of a ride back to Denver where he had left his BT-9. Jack was an instructor pilot at Randolph Field, and on his cross-country to Salt Lake the previous week, had encountered bad weather, and as a result, had left his aircraft there and come the rest of the way by land transit

I was due a cross-country of my own and was able to obtain the use of our squadron A-17. Our bomb group, the 7th, had just recently moved from Hamilton Field, near San Francisco, to Salt Lake City. And our home was actually at Fort Douglas, a beautiful setting against the mountains on the east side of Salt Lake City. The commercial airport at Salt Lake City was almost totally taken over by our group and its array of B-17Bs, Cs and Ds, and some B-18s and a smattering of other small types (of which our A-17 was one).

And so it was that Jack and I climbed into the A-17 and took off for Denver. It was a morning flight and we were forecast to have clear weather all the way. The A-17 was a strongly built aircraft, designed for low altitude "attack" work. As a result its engine was not supercharged for higher altitude flight. In fact, the flight level for instrument work between Salt Lake and Denver was necessarily in excess of 14,000 ft., since the mountain tops went that high. However, in clear weather one could fly along the light line and make out very nicely at 8,000 ft. For those who may never have heard of a light line, since they have long-since been removed, the light line was made up of a series of towers, each within reasonable sight of those on either side, from which a beacon would shine, emitting a coded signal for identification. A pilot could expect to always have one and perhaps two in sight at any one time, and thus be guided along the route. Of course the route selected was always that which enabled flight at the lowest reasonable altitude. Thus when going through hilly or mountainous terrain the route necessarily resulted in one wending his way around mountains and flying through valleys – a most pleasant and enjoyable experience in clear weather, but hazardous in low clouds and poor visibility.

The A-17 with its unsupercharged engine could barely manage to produce enough power to yield an airspeed of 100 to 105 mph at 8,000 ft., but since the weather was clear this was not a problem. The flight to Denver, although slow and tedious (3:25 time logged), was completed without incident.

After bidding Jack good bye, eating a late lunch, gettin the aircraft serviced with fuel and filing my flight plan for th return flight to Salt Lake, I was ready to take off. My clearanc showed that I was to have clear weather all the way back Salt Lake City. Not to my liking, we would be in darknes well before reaching our destination, however, since th forecast was CAVU all the way, I gave it no thought. Ceilin and visibility unlimited is always a pleasant sound to a pil about to embark on a flight.

The take-off at Denver's Lowery Field required a goo portion of the rather long runway. The field elevation as recall was approximately one mile, or something over 5,00 ft. This elevation of course had its effect on my unsupercharge engine and accounted for the longer take-off run.

After becoming airborne the aircraft gradually regaine the 8,000 ft. cruising altitude, and we were off on our wa back to Salt Lake City Airport. After a couple of hou darkness began to fall, and the light-line beacons could mor easily be picked out. It required very little course correctio to follow the lights during this time, since the more moun tainous portion of the trip was that portion just before gettin into Salt Lake City. In due time I was able to identify th auxiliary landing field at Fort Bridger, and a little later th one at Green River. Both were outlined with lights and eac had a green beacon in its light-line tower. The en route ligh line beacons were red, with the green signifying a landin facility.

Shortly after passing the Green River auxiliary field, began to encounter wisps of clouds at my altitude of 8,000 f As I progressed, the cloudiness became more pronounce Darkness had set in and I was quickly becoming of the min that I would soon be obliged to take some corrective actio The weather was not such that I could expect to continu under visual flight rules. I was aware that the radio beam fo instrument use was to my north. My plan of action was t increase altitude to at least 10,000 ft., turn to a northerl heading, intercept the beam, then proceed east on the bean until reaching clear weather, then find and land at eithe Green River or Fort Bridger auxiliary field.

However, as I made my turn to the north (now o instruments) I noticed that the needle in the needle/bal instrument did not move. So I straightened up then tried turn in the opposite direction - still no movement. Then checked the vacuum, which seemed normal. But neverthe less, I chose to cross-check it, so I switched to the Ventur vacuum. The gauge showed less, as to be expected, than th engine pump was delivering. So I switched back to th engine pump. Vacuum was not the problem. Obviously th flight instrument was faulty. I had known from the outset o the flight that the artificial horizon was not functional, bu this was often to be expected in that era. So my usefu instruments included airspeed, rate of climb, altimeter, com pass and directional gyro, and of course, the ball without it needle.

I was able to worry the aircraft up to 10,000 ft., and als to intersect the beam, and to get myself headed in an easterl direction. It did not occur to me that in increasing altitude had probably gotten myself into more cloud cover than tha which I had experienced at the lower altitude. Thus afte flying for a time when I felt that I should have been in the

:lear, I was still on instruments. However, as I flew along, I began to see breaks in the clouds in that I could sometimes catch glimpses of lights on the ground. Since I was trying to locate one of the auxiliary fields I elected to take a look when one of the breaks occurred. After taking a look at one such break, I found myself in a spiral. I corrected by kicking top rudder and easing back on the stick, then correcting back to the easterly heading of the beam. I made several attempts to locate myself and one of the auxiliaries. Each time I wound up in a spiral, always to the right. Each time the corrective procedure was the same. However, on the last such maneuver I was to make in that aircraft, I noticed that as I eased back on the stick the engine RPM increased!

I immediately concluded that I was beyond the vertical - that I was closer to being upside down than I was to being right side up. The A-17 was equipped with a two position propeller - a relatively flat pitch to be used for take-off and climbing, and a steeper pitch for cruising. With the increase in RPM it was clear that the aircraft was heading more steeply down as I eased back on the stick. My immediate conclusion was that if I could not fly the aircraft on instruments while it was right side up, I sure as hell could not expect to fly it on instruments while it was upside down. With that I made preparation to bail out.

I immediately pulled the hatch back, unbuckled the seat belt, and started to roll out the left side of the cockpit. But as I looked back and saw the horizontal fin which stuck out some ten or twelve feet – I knew that if I rolled out I would surely be hit by the fin and perhaps be seriously injured. So I pulled back into the cockpit and then elected to dive for the left wingtip light. I had once put in a stint as a life guard and had been taught how to make a long and purposeful dive to get to someone in trouble quickly. I made such a dive for that little red light.

The feeling of being in space reminded me of being on the softest possible feather bed. I brought myself out of this reverie with the thought that I should pull the ripcord. Which I did, but I apparently did not pull it hard enough, since nothing happened. So I pulled it again. This time the entire ripcord cable and all came out. The chute popped open. My next thought was that the boys always like to kid around that the person that parachutes is always so excited that they lose the ripcord. So I carefully transferred it to my left hand and was so holding it when I suddenly hit the ground. With the darkness and the cloudiness, I had no glimpse of the ground before striking it. Fortunately I landed on soft ground but was so relaxed that I collapsed completely, with my head coming down and my left cheekbone near the eye striking the ripcord ring which I was so carefully holding in my left hand. A black eye – my only injury.

When I left the aircraft I neglected to pull the throttle back to cut the magnetos, so the aircraft kept flying for awhile. While floating in the chute I heard the aircraft go through a couple of climb/dive sequences. At one point I thought it might hit me or at least catch my chute, it seemed so close. As it turned out, the aircraft's wheels caught on the top of a ridge, the engine was immediately torn out and rolled to the bottom of the canyon, several hundred yards from the aircraft. The aircraft itself flipped over on its back and slid down the hill a short distance. The fuselage construction was so strong that it did not seem to be distorted or bent up at all. (A week later I hiked up to see for myself).

I had no way of knowing how close or far away the aircraft was from where I landed. After I picked myself up I took stock of my situation. I knew only that I was somewhere in the Rockies, near the top of a ridge. My clothing consisted of a light flight suit over my pinks or greens trousers and shirt with French Shriner oxfords. I elected to carry my chute along. I had a pen light which afforded some help. But seeing anything was difficult. I landed about 9 P.M. and hiked until about midnight. By then I had worked up considerable body heat and when I wrapped up in the chute I was able to sleep for a while. I was awakened by being cold about five or six o'clock in the morning. Dawn was just breaking.

With daylight I discovered that there was a stream between the ridge I was on and an adjoining one. My direction of walking the night before was consistent with the direction the stream seemed to follow. I decided to follow the stream on the basis that it would sooner or later bring me to civilization. At first I thought that I would be able to walk beside the stream, but soon found that in places the slopes to the stream were so steep that I had to get down in the stream itself to walk. Before doing this I provided myself with two sapling staffs for two reasons. I could establish how deep the stream was before I took the next step and they were an excellent aid to keep balance. Somewhere, in my involvement with the stream I lost the ripcord. I had carefully placed it in a lower leg pocket. But the pocket had no closure and I may have lost my footing sufficiently to have permitted it to fall out. In any event it was no longer with me when I looked for it.

The ridges finally opened up enough to where I could walk along side instead of in the stream. The first sign of civilization I encountered was a trap with a porcupine doing his best to get loose. I elected to leave well enough alone and kept going. The next thing was a trappers abode with a pen for horses - but no occupants. Then I heard a train whistle, so I knew that I was going in the right direction. At least I could follow a train track until it brought me to civilization. Then the ground flattened out completely. Thus I could walk faster which I did. By about 8 A.M. I found myself at a paved highway. I immediately got out on it and started walking in a westerly direction. I had not gone more than a few hundred yards before a fellow came along in a Model-A Ford, kind of looking as he drove along at a moderate speed. When he got to me he stopped and asked me if I was the pilot they all had been out looking for the night before. I allowed I must be. He took me back to his place which was a mile or two down the road in the direction I had chosen to walk. He was William Kyes from a place known as Taggart's Canyon Camp. He called the highway patrol and reported my presence and I called the air base.

They were very nice people and they invited me to share breakfast with them. I recall that the breakfast consisted of eggs and bacon, toast and probably home fries. When they passed the platter to me I apparently helped myself to the entire fare. At least that is what I was told later. In any event they were fine people and they said that the air base people had been out the night before with several cars, trucks, ambulances, driving up and down the highway flashing lights up the mountainsides looking for any sight of me or the aircraft.

The place that I went down was known as Devil's Slide. It is just a few miles east of Ogden, Utah.

The tragic epilogue to all of this is that in the course of the following few days several individuals hiked up to see the remains of the aircraft. One poor soul died in the process of hiking there.

# TARGET: COLOGNE

*by Emil M. Klym*

During the Battle of the Bulge, our group's mission was to fix over the German troops at an altitude of 100 ft., strafe and drop 100 lb. anti-personnel bombs if the German army advanced beyond a designated point in their drive to take Antwerp, Belgium. We were put on standby and spent most of the day on the flight line awaiting the order to take off. The flight crews were very quiet and kept their thoughts to themselves until the news was received that the mission was canceled. At that time they broke into smiles and there was laughter and chattering. I have often wondered what the outcome would have been and how many of us would have not returned if the mission had not been canceled.

The 323rd B.G. (M) mission on 10 February 1945, was to bomb a target in Cologne, Germany. As we approached the target the flak was extremely heavy and somewhere on the bomb run our aircraft took a hard hit. The right engine was knocked out, the right windshield was blown out and the bomb bay doors damaged and would not close electrically. When the pilot informed me of the problem, I went to the bomb bay, closed the doors with the manual crank system and then went to the cockpit to assess our predicament. What I saw was all bad. We were on our own, limping on one engine (the formation had left us as we could not keep up with them on one engine) the navigator was not sure of our whereabouts, we were losing altitude and running low on fuel. It was evident that we were all going to bail out soon and shortly after we had received a tremendous shelling from a town.

Just off our right wing the pilot, Capt. William r. Adams, from Long Island, NY, instructed the crew to bail out. I opened the bomb bay doors and all of the crew, except the pilot and I, jumped without any hesitation on their part. When I returned and informed the pilot that all of the crew had left, he instructed me to jump. I jumped and after my chute opened I kept counting the chutes and watching the plane and praying that I would see the pilot's chute open.

The plane made a half circle to my left and burst into flames when it hit the ground. At that time, I looked down and saw the trees directly below me. I covered my face with my arms and fell into the trees. The next thing I knew I was lying on the ground, flat on my back, entangled and covered with my chute, and unable to move or utter a sound. I lay there for some time before some Belgium people found me and carried me to a farm house where all of the crew, except the pilot, had gathered.

When I finally recovered my voice I told the crew that the pilot must have gone down with the plane as I did not see his chute open, and we were all elated when he walked in a few

minutes later. He told us that the plane's altimeter was down to 1,500 ft. when he jumped.

Since we had just gotten over into our side of the front lines, the pilot was able to locate a British ambulance that took me to a hospital in Brussels, Belgium. After hospital stops in France and England, I eventually arrived in the States where I spent seven months recovering from a broken back.

# BAILOUT OVER CHINA

*Source: Sept. 1987 issue of "Airman" by CMSgt. Vickie M. Graham*

Pete Kouzes was a radio operator on a mission carrying a load of 100-octane aviation gas from Luliang, China, about 60 miles east of Kunming, to a place called Lahoko on Dec. 10, 1944.

Staff sergeant Kouzes' flying partners that day included 2nd Lt. Albert J. Fisher, pilot; 1st Lt. George (Mac) McGuire, co-pilot; and Sgt. Elon (Pat) Patterson, flight engineer on the C-46 Commando. Their job was to ferry critical war supplies over the treacherous "Hump" in China.

The crew got lost and as they neared what they thought to be their destination ground orders were to descend below the cloud cover. No familiar landmarks were to be seen.

Three Japanese fighter planes and one biplane were on them. The fighters missed the first pass but the plane took a direct hit in the left engine the next time around.

The pilot ordered the crew to bail out. Kouzes managed to get a "Mayday" call out before heading aft to don his chute. On the third pass, bullets pierced the fuselage and 55-gallon drums, spilling the high-octane gas everywhere.

By the time Kouzes got to his chute it was soaked with the highly flammable gas. The airman was scared and prayed then jumped out the cargo door. He didn't even wait the three seconds before pulling the ripcord.

The plane turned 180 degrees and exploded only seconds after the crew had made their exit. Burning metal spewed over the descending men and they wondered if the chutes would catch fire.

Kouzes was not out of danger yet. Out of nowhere one of the enemy planes headed directly towards him and he could see the sparks from its wingmounted guns. All 22 years of his life flashed quickly before him and the young man prayed again. Inspiration hit and he fell limp in the chute, pretending to be dead. The ploy worked and the Japanese fighter banked to the left and disappeared.

Three of the four crew members survived. Ground witnesses said that Mac's chute had been peppered with bullet holes and he probably died when he hit the ground.

The landing site was controlled by a Chinese warlord. An interpreter arrived three days later with a dictionary to help communications.

The men were taken to the warlord's fortified village and stayed there until two days before Christmas. The Chinese communists attacked the warlord's hold on Dec. 23rd.

Kouzes later said they were up in an attic and mortars were going off all around them. Sgt. Patterson tossed his American flight jacket out of the window and the fighting stopped long enough for the communists (who were anti-Japanese) to get them out.

The three survivors eventually met up with a downed P-51 pilot and a B-29 crew and 84 days after their bailout were finally reunited with their squadron. Hundreds of miles had been crossed by oxcart over rugged country trails to evade capture in the Japanese-held territory. The men had many interesting experiences along the way, including being entertained at a banquet in their honor (for propaganda purposes) during the holidays. Along the trail they often played softball to pass the time.

## THE CATERPILLAR MEMBERSHIP COMMITTEE

### by Frank Ramsey

During World War II, I served as a crew member of a B-24 Liberator bomber. I was assigned to the 776th Squadron, 464th Heavy Bombardment Group, 15th Air Force, serving in Italy. On my 39th mission, we were to hit the oil field of Ploesti, Rumania. This was my third trip to Ploesti.

As we neared our target, we began to experience engine trouble on two of our engines. Rather than fall out of formation and become easy prey for enemy fighters, we chose to increase the manifold pressure on the other two engines and stay in formation and drop our bombs on the target. All went as well as could be expected until we reached Yugoslavia. We were unable to maintain our altitude. We descended to 5,000 feet and there were mountains in the area 6,000 feet. The sky was overcast and visibility was poor. Rather than risk slamming into a mountain peak, the pilot, Captain Bruce C. Cater, gave the order to bail out.

After a couple of days in the mountains of Yugoslavia, I was able to unite with four other members of our ten man crew. We had bailed out in a rugged section of the Kapela mountain range. The people in this section were very poor but were willing to share their meager supply of food with us. The food consisted mostly of goat cheese and bread baked on a hearth by putting hot ashes and coals on it.

We traveled by night and hid out by day. In this section of Yugoslavia, there were several factions fighting each other. There were the Germans, Partisans, Chetniks, Bulgars and others who I am now unable to recall. We never really knew who we could trust.

There was one man who claimed to be affiliated with the Partisans. He acted as our guide for the 39 days we spent in Yugoslavia. During those 39 days, we must have hiked some 250 or 300 miles, and a good part of it was done by night. By the time we had reached our contact point, we had been joined by a five man English air crew and parts of two other American crews, bringing our total up to 25 airmen.

Our contact spot was on open field, lighted by brush fire, that was to be used as a landing strip for a Russian crew flying an American C-47. The three man Russian crew landed the plane without incident and by now there were 25 to 30 wounded Partisans who were to be loaded along with us and flown back to the hospital in Bari, Italy. There we learned that the other five members of our crew, who had been aided by the Chetniks, had gotten back to base 12 days earlier.

In July, 1986, six members of our "bail-out crew" got together at the 464th Bomb Group reunion that was held in Manitowac, WI.

## WHY BURN THE STARS AND STRIPES?

### Submitted by Johhny Brown

Why burn me? What have I done to be put to death? Yes, I have led people into battle. I did not order them there, but I went to guide, support, and represent them. I have been held by one of my children who marched into the fire, the destruction, and the very hell of war. Above the death and smoke, I have stood tall, flying as an ensign for those who fought against tyranny and hatred.

I flew when the British attacked, all night I flew, marking where those who fought for freedom stood. And when the morning came, I was still there. I was at Gettysburg, where my sons fought each other hand-to-hand; senseless hatred fueling them on. I still fly there, marking the site as a shrine and reminding the living their death was not in vain.

I still fly at Pearl Harbor, and not at half-mast. There, an enemy thought my children would falter. But they were wrong. Today, the children of the dead come to look and be reminded that these heroes died that freedom might live on.

My sister with the torch has held her vigil, watching over the ocean for our children to return from war. I was there, flying from the highest mast, telling her, "Here are our children, here are those who fought for the freedom of mankind."

I stood erect and proud, representing all who love freedom, as my children who were prisoners of war were returned to their loved ones.

I have covered the bodies of those who gave their lives that we might be free. As they were returned to grieving families, I was a reminder of the freedom and truths which their children died to uphold.

But I have also led in peace. My stripes are a reminder to the world that they can in fact live in peace as separate, yet United States. My stars are a new constellation in the sky, guiding all who yearn for freedom and equality to these shores.

Missionaries, peace workers, and other ambassadors carry me around the world, and wherever I go, people follow; they know those who bear me also bear that which I stand for.

What do I stand for? How can you ask, you who have lived so long under my stars?

I am a declaration of independence. I am independence from fear, from tyranny, from slavery. I am a visual reminder that "...all men are created equal..."

I represent a constitution; a most unique document. It is a contract, a covenant, wherein the people allow a few to govern the many.

I stand for the Bill of Rights that gave those people a margin of civil liberty still unknown to the rest of the world. I stand for the very liberty that allows you to burn me.

I stand for a government that is, in turn, governed by the people; a government that is of the people, by the people, and for the people. It is a government that must listen to, and obey the people.

I stand for the people, and in doing so I stand with the people. I stand over the parks and playgrounds, sites where children can play free from Cossacks on playgrounds or slave-drivers searching for new stock.

I fly above the offices, factories, farms, and shops of the nations, reminding the world that here mankind is free to buy or sell, to spend or save, and to become all they can achieve by the sweat of their own brow.

And, I fly at our ports. Those who lived with war, with hatred, with oppression, see me as a sign of a better life and flock with their children to the light of my shadow. They join my family and the circle of freedom grows larger.

In every American heart beats the song of freedom. But sometimes the child does not understand the words. The American parent then points into the sky, where I unfold in the breeze, and tells the child that I will always lead them, that where I go they will find freedom and dignity.

I am freedom. I am the heart of every American. I am the heart and soul of America.

When tyrants see my children burning me they smile, for they see the heart and soul of America dying. Why burn me?

## EVADING THE ENEMY

*by W.C. Dubose*

It was about 3:00 p.m. on Saturday, June 17, 1944. We were flying over the beautiful green wheat fields and farms of northern France. Our mission that day was to glide two eleven second delayed action thousand pound bombs into a railroad bridge across the Somme River near Perrone. The purpose was to help cut the German supply lines leading to Normandy.

There were 48 of us on this mission. We were to take turns flying in elements of two about 100 feet above the water, release our bombs just as the target passed through our gun sight, pass over the bridge and make a sharp climbing turn to the left. Not far beyond the bridge, the river swept left and on the right bank was a German airfield.

When our turn came, my flight leader, Captain Don Penn and I dove down and flew along the river toward the bridge. We could see French people atop the high banks waving and cheering us on.

Apparently we were only about 10 seconds behind the aircraft ahead of us. As we lined up on the bridge, one of the bombs from the previous P-38 went off in the water far short of the bridge and I could not help flying through the splash.

Don released one bomb instead of two so I quickly changed my trigger so I would also only release one bomb. My thought at that moment was that we were to make a second run. (Interesting how an incorrect assumption can change your whole life. No, I didn't go on to become a famous ace or pursue a flying career; maybe that's why I am alive and well today.)

Don then advised our leader that one of his bombs had hung-up and would try to jettison it. He did this on a freight train some 10 to 15 miles away and blew the hell out of it. I asked Don what I should do with my extra bomb. "Drop it on a target of opportunity," was his answer. I saw one sitting at the station of a very small village I later learned was called Chaulnes. What an opportunity! Zip along just above the train and drop a thousand pound bomb into the cab. This was going to be the epitome of all boiler explosions. What fun for an immature kid who had just turned 20.

Things went as planned for the first few seconds as I zipped along about 20 feet above the train. Then all hell broke loose. German flack positions on both sides of the track opened up on me. They couldn't miss. Shells flew through my wings and nacelles. Instantly, I triggered my guns and dropped the bomb, but it was too late. My plane was on fire and Don was screaming for me to bail out.

I pulled up, jettisoned the canopy, unhooked my seat belt and decided to go out over the top because smoke was pouring into the cockpit. After I pushed myself up into the slip stream, I was pinned against the back of the canopy hanging half in and out of the plane. I couldn't move and just dangled there for a few seconds until my plane turned over into the dead engine and started down. A few seconds later I was pulled free. I saw the tail whip by and pulled the rip cord. My parachute opened with an explosion and I saw my plane burning on the ground. I looked around to see where I was going to hit and tried to turn my chute so I would hit facing foreword but it was too late.

I slammed into a wheat field going sideways and broke and dislocated my right ankle and sprained my left. I hit with such force that my one-man dingy was popped loose and spread all over the area. I pulled in my collapsed parachute and unhooked my harness. We had been told to bury it, but I did not have the strength, so left it there and crawled about 30 feet to a dirt road.

At the top of a hedge row on the other side I could see German soldiers running from the village of Chaulnes toward my plane which had hit about 100 yards from where I had landed.

I crawled back across the road into the wheat field and headed toward some trees about a mile away. As I crawled toward the edge of the field, I saw two German soldiers running toward me. I thought they saw me so I laid down. They got to the edge of the wheat field and turned toward the road. Had they run straight ahead, they would have tripped over me.

At this point, I decided it was too risky to crawl across an open field, so I headed back to the hedge row on the side of the road and hid myself with leaves and bushes. I looked at my watch; it was 3:15 p.m.

As I lay in that hedge row, 40 German soldiers combed the area looking for me. A few came within five feet of me as they walked and drove along the road. Apparently they gave up the search after a few hours. As evening approached, I could see the lights of a farm house several miles away. I decided that would become my destination. Perhaps the people who lived there would help me.

It did not get dark until after 10:00 p.m. (double daylight saving time). I decided it was safe to start crawling through the wheat fields toward the lights of that farm house. I hadn't crawled very far before I realized this was going to be a painful task. My green summer flying suit gave little protection to my knees.

Perhaps if I got to my feet and could find something to use as a cane, I could hobble to my destination. I saw a concrete power pole in the distance. I crawled to it and tried to pull myself to my feet. It was too painful. I could not put any weight on my ankles. I sat down, tore the legs from my flight suit and wrapped them around my knees. This gave me some relief.

As I crawled through the wheat fields, I tried to camouflage my path by crawling in and out of the fields in places that would be difficult for some one to follow. I was told later that some French people did follow my route but it was not easy.

Late that night, I crawled to a spot where my hand reached out into empty space. I stopped and felt around for solid ground. There was none. Then I heard what sounded like German voices down below me. I slowly and cautiously backed away. Had I crawled up to the edge of a German flack position? I will never know.

As the early morning light appeared, I could see that straight ahead of me was the railroad line. It's bed was higher than the wheat so if I tried to crawl over it in the light, I could be seen for hundreds of yards in all directions. There went my plan to reach the farmhouse. Just as well, it still seemed to be several miles away.

My second plan was to turn left and head toward the village of Chaulnes. Perhaps I could find some friendly French people who would hide me.

By this time my knees were bleeding and every movement was painful. I decided to sit on my back-side and push myself along backwards using my hands to propel me. I did this from early morning until late that afternoon.

This area had been a major battlefield during World War I. As I pushed myself along that day, I came across a broken saber and a rusty hand grenade from that war. I would liked to have kept them as souvenirs but that just was not practical.

About 5:00 p.m., I reached a dirt road near the outskirts of Chaulnes. I sat in the wheat at the edge of the road. Finally, a woman and her young daughter walked by. I got to my knees so they could see me and yelled, "I am an American."

The woman was startled but kept calm. She grabbed her daughter's hand and pulled her along, apparently telling her not to look back at me or say anything.

Approximately 30 minutes later a man came walking down the road in my direction and seemed to be looking for me. I hollered at him and he came over, knelt down and spoke to me.

Using my French-English translation card from my escape kit, I was able to show him sentences that stated that I was an American airman, shot down, injured, thirsty, hungry and wanted to be hid. He was cautious and asked if I could speak German, Spanish, or French. I was not able to speak any of these languages, but from the little I had learned in high school, I could understand what he was getting at. He wanted to make sure I was not a German spy planted there to find out who were members of the French Resistance. He motioned for me to stay low in the wheat so I would not be seen; then he left.

Sometime later, two teenage boys came looking for me. One motioned for me to crawl across the road and follow them. I got across the road but was not able to crawl any further. One of the boys pulled me onto his back and ran about 100 yards down the road to a driveway leading into a farm. His father was waiting for us with a wooden wheelbarrow. They put me in it, put an old piece of carpeting over me, and wheeled me back to the barn behind the house. In the corner of the barn was a pile of grain. Somehow, they pushed me up and over the crest so I was hidden between the grain and the wall. I quickly fell asleep.

Hours later when it was dark, I heard voices below and could tell someone was crawling up toward me. They brought me a piece of bread and a bottle of water. After I had eaten, they pulled me down and took me into the house. I was taken to a bedroom on the second floor where we attempted to communicate.

I do not recall how many people were there, but I do know they were concerned with my physical condition. One lady was from the local Red Cross. They called her "Mademoiselle Rouge." There was nothing she or anyone else could do to help my ankle.

They took what was left of my flight suit, my GI pants, and shirt and gave me a sweat shirt and pair of pants to wear. I kept my wings and dog tags.

The next morning I was awakened and a man who could speak English appeared. He was a former World War I English soldier who had married a French lady and settled in France. They put me in his horse-drawn carriage, told me to lay down so I was hidden. I was then driven about ten miles out into the country to his farm.

He told me that their big beautiful home had been burned by the German occupation forces as they invaded France. They now lived in the servants quarters nearby. There was not room for me in this one bedroom house which they shared with their daughter. They had cleaned out a chicken coop for me.

Actually, the chicken coop was adequate. It was clean, had a cot and an end table; the chickens next door were noisy, but good company.

While there, I was made to stand and hobble around with a cane as soon as the swelling went down. I am sure they were concerned about getting me mobile as soon as possible and moved somewhere else. Both my legs were black and blue all the way to my hips. I had really hit the ground hard!

Two weeks later, I was transported back to the village of Chaulnes and put into the care of the Edward Leblanc family. They lived on the main street about two blocks from the railroad station where I had tried to bomb that train. The downstairs front of their house was a store. Their living quarters were behind the store and upstairs. I lived in a bedroom on the second floor.

During my approximately three weeks stay with the Leblancs, the railroad yard was bombed by B-26 bombers. I was watching them fly over and saw and heard the bombs falling. Even with my bad ankles, I ran down the stairs and out the back door to their slit trench bomb shelter. A piece of metal from one of the bombs went through the wall of my room. Numerous French civilians and German soldiers were killed or wounded.

The French had given me false identification papers and used the picture from my escape kit (we all carried our picture taken in civilian clothes—just in case we were shot down). My French name was Jean Pierre Dubose. I was supposed to be a friend of the family from the Normandy area who was deaf and mute—injured during the invasion of France.

The Leblancs had neighbors sympathetic to the Russians who visited quite often unannounced. They would bring their map of Europe and discuss the latest positions of the Russian and American fronts. The Leblancs were not sure their friends should know they were hiding an American flier so every time they visited, I played the roll of a deaf and mute person.

Papa Leblanc was always testing me to make sure my act was convincing. One evening we were seated at the dining room table and my back was to the door. The Russian family dropped in to discuss the latest position of the fronts. Papa walked in the door and dropped an empty metal bucket on the hard floor just behind me. The noise startled everyone. Luckily, I did not flinch an eyebrow or muscle.

They had a daughter Suzanne who was married, but her husband was being held as a prisoner in Germany. The Leblancs also had a son somewhere in North Africa with what was left of the French navy.

He had left his parents a Navy telescope which I used everyday to spy on the German flack positions. One day as I was watching, the German troops were lined up and an officer was addressing them. He gave several of them medals—apparently for shooting me down!

The French insisted that I get out of the house occasionally to get some exercise. They got me a bicycle and we rode along the dirt road beside the flack positions so I could see my enemy. They also took me out to the hole in the ground where my P-38 had hit. I can assure you, I was not enthusiastic about making these bicycle trips.

It was more interesting to watch the 100 regular German troops and 100 SS troops, who occupied this village and manned the flack positions, march down the main street every evening singing German songs. Apparently they did this to keep up their morale and show the French who was boss.

The day after the attempt to assassinate Hitler, there was a lot of confusion among the troops and some of the officers left town. The soldiers milled around the village discussing what they should do.

I stood at an open window upstairs in the back of the Leblancs' house and watched three or four German soldiers standing on a road just behind their yard. They saw me watching and yelled something. I had no idea what they were saying so I just stood there. One of the soldiers pulled out his pistol and fired at me. I got the point and moved away from the window.

All the French families tried to feed me as best they could. I ate a lot of boiled tripe, strawberry sandwiches and chicken. The sanitary conditions were not the best and a few days before I was to be moved to another family, I got dysentery. It made me very sick and very weak. However, the French schoolteacher who was to take care of me for the next few weeks showed-up and we took off on our bikes. I was so weak, I had to walk beside my bicycle and push it up any kind of incline.

On one of these walks up a hill, we were accompanied by a German soldier. My French schoolteacher friend and he talked all the way up the hill. I just played deaf and mute and was scared.

I lived with this family about two weeks and was told that I would be transported by car to another location. By the way, every move I made was north—the opposite of the direction I wanted to go.

This time I was picked up by two members of the French Resistance. In the car was an American B-17 gunner—my first contact in a long time with someone with whom I cou really communicate.

As we drove north through the countryside we saw group of P-47s strafing. Our driver immediately pulled in a farmyard and parked under some trees. We all ran for th house. Shortly after we got into the house, two truck loads German soldiers drove into the yard and parked under th same trees. As the soldiers ran for the house, my new B-1 gunner friend and I were told to run out the back and hid behind an outhouse near the edge of a field. We hid there fo what seemed like eternity, watching the P-47s strafe on on side of us and the German soldiers milling around in th house on the other. Finally, the P-47s left, the Germans lef and we departed.

We were taken to the apartment of a Madame Heller i the village of Billy-Montigny. She was the head of the Frenc Resistance in that area. She was Australian and her husban a Hungarian photographer. They had been caught in Franc when the Germans invaded France. Neither were Frenc citizens but were forced to stay there during the war.

We arrived in time for dinner and were told we would b spending the night with a brewer on the edge of town. W were told to walk down the stairs to the street, turn right, g to the corner, cross the street and wait for a car to pick us up We did this but no car showed up. We waited and waited. I was getting late. It was about 8:45 p.m. and curfew started a 9:00 p.m.

Finally, a young man riding a bicycle showed up an motioned for us to follow him. He could see that I coul hardly walk so he put me on the handlebars and drov several blocks. He told me to get off and walk straight ahead He then went back to get my friend and in this manne shuttled us to the Brewers home on the edge of town. W were lucky no German soldiers were around or we woul have been apprehended for being out after curfew.

Our new French host offered us food and drink. I was to scared to be hungry. The next morning, we were awakene and given coffee and cognac.

Before the occupation, he ran a brewery next to his home He was proud of all the things he had been able to hide fron the Germans. He took us for a tour of his brewery and property.

Under the loading dock of his brewery was a secre entrance to an area where he had stacked hundreds of case of champaign and a room where he kept his radio. Here h could listen to broadcasts from the BBC directed to the peopl of the French Resistance. Radios were outlawed by the Ger mans. Upstairs in one corner of his brewery, were stacke hundreds of cases of empty bottles. His car was hidder behind them.

The large garden of his home was beautifully land scaped. In the center he had poured a concrete slab. Next to it lay an unexploded 1,000 pound bomb with the detonato still attached. He was going to stand it on the slab as souvenir of the war. Next to his garage, in a wood pile, he ha stored two British 200 pound unexploded bombs. Needles to say, we were not thrilled with his tour.

Madame Heller and her driver came by that afternoon t take us to our next destination. On the way, we picked up typical looking Englishman, mustache and all. Madame

Heller tucked our wings and dog tags in her bosom. My American friend and I sat on either side of the Englishman and asked him to look the other way whenever we passed any German soldiers.

We drove down one narrow dirt road beside hundreds of marching German soldiers. My friend and I made sure the Englishman was looking the other way. We came to a gate guarded by several German soldiers. The soldiers, the driver, and Madame Heller spoke for a few minutes. Finally the guards raised the gate and we drove through. My heart was in my mouth!

Our destination was the home of the Mr. and Mrs. Dernancourt who lived in the city of Lens. Six other allied airmen already lived with them. From that point on, the nine of us lived together in the rooms above their store which was located on the main street of this city.

By day, we played poker with our "escape money," sunned ourselves in the brick courtyard behind their house, shared by a pig, or watched the activity of the German soldiers from the windows upstairs. We all picked out pretty girls on the street whom we would like to meet when we were liberated. We also helped prepare the meals.

We ate a lot of soup which contained everything our hosts and their close friends could conjure up to put into the pot. We spent many hours grinding, peeling and stirring.

This is where I met Clifford O. Williams, a P-38 pilot from the 343rd Fighter Squadron. He had been shot down the same day that I got into the 55th Fighter Group. The nine of us consisted of two Australians, two Canadians, two Englishmen and three Americans. Madame Heller had also found places for about 14 other Allied airmen to live in the Billy-Montigny, Lens area of northern France.

One evening while all of us were sitting around the dining room tables eating, we heard a scream from the young teenage girl who was minding the store up front. She came running back to warn us that a German truck with many soldiers had driven up and parked in front of the store. We all thought that someone had turned us in and we would be taken prisoners and the French people shot.

As it turned out, the Germans had stopped to take hostages. They made it a point to take a husband or wife from a family—never both. They were put in the truck and taken off to be shot. This was shortly before we were liberated by the English 1st Army. The French were out every night blowing up bridges, killing German troops and cutting telephone wires. For several weeks, we could hear the explosions, see the flashes of light as the French Resistance did their thing—things like putting unexploded bombs onto carts and dragging them under railroad bridges where they hoped they would eventually explode.

It was interesting to watch the German troops retreat. For several days before the English 1st Army arrived, the Germans came down the main street heading north in every conceivable mode of transportation imaginable—trucks, cars, bicycles, horses, horse drawn carts, tanks, on foot, etc. This went on 24 hours a day.

Finally, the English 1st Army rumbled through with tanks and trucks and many troops. Everyone was out waving and cheering. Flags were flying and people were crying tears of joy. As the troops sped by, they threw us cigarettes and candy bars but would not stop.

When a convoy did stop, we spoke to them and they got word to their officers that we were allied airmen and needed transportation back to Paris. This was arranged but it took a few days.

In the meantime, we were treated as heros by the townspeople, were given a banquet and asked to march in a liberation parade. In the parade with us were French collaborators. The women had their hair shaved and they were kicked and spit upon as they marched along. The pretty girl I had picked to meet after liberation was one of those women!

Several days later, several of us were put in a truck and taken north to the front lines where we were turned over to the American 1st Army. We spent one night in a German prison compound. It was just behind the lines and a place where the captured Germans were brought before they were sent south to prison camps. They were some tough looking SS troops among them.

The next day we were put in a truck along with other Allied airmen who had been found along the way and driven back to Paris. There we were taken to the Hotel Maurice. This had been the German Army Headquarters in Paris during the occupation but was now a place where all the evadees and escapees were brought to be interrogated. We spent two or three days here. Our false identification was taken and we were given a GI shirt, pants, pair of shoes and socks. I took a cold water bath, the first bath in three months.

Paris still had pockets of Germans but that did not bother us. As soon as possible, we were in the sidewalk cafes drinking champaign and trying to pick-up French girls.

Then we were flown by C-47 back to London. I was put into a tent hospital north of London where I spent another month or so. Every V-1 shot toward London seemed to pass over this location. It was another terrifying experience. I was flown back to the good old United States in November of 1944 and back to my home state of California. Again I was put in a hospital where they tried to repair my ankle.

And you know, the irony of all this is that very few people understand why I like to watch World War II Air Force movies on TV and will not buy a Japanese or German car!

Viva La France and God Bless America!

## THE FRENCH UNDERGROUND
*by William E. Martin "Bill"*
*8th AF, 384th BG, 547th Bomb Sqdn.*

Our crew(#73) led the squadron on the strike against Schweinfurt on October 14, 1943. This was our 21st mission against German targets. We were well on our way to completing the 25 missions required and were confident we would now complete the remaining five and be relieved of combat duty. At this point, only one crew, the Memphis Bell, had reached this goal.

Suddenly Folk Wolf 190s were attacking the remnants of our squadron from 1 o'clock high. There was an explosion of 20 mill shells inside the plane—one directly over the bull turret near my post as left waist gunner. The right wing burst into flame and fire was sweeping past the right waist window—there was a cry "BAIL OUT!" The fire at the waist door prevented jumping out there. The ball turret gunner, waist gunner, the tail gunner, and I all rolled out the tail gunner's escape hatch. I searched frantically for the release on my chest type chute, finally finding it on the wrong side (I had snapped it on backwards!) Then a Folk Wolf buzzed me and I thought the chute was going to collapse.

I landed in a plowed field and the waiting arms of a French farmer (Henri Pierre, Courcalles, France). He rolled up my chute. I gave him my escape kit and chute harness and he pointed to the nearby woods. I hid in the woods that day and night and watched German troops search for downed airmen. The villagers outfitted me and Francis Sylvia (another crew member who landed nearby) in cast off clothing and hid us in the farmhouse. Members of the French Underground arrived and verified that we were indeed American airmen and made plans for our escape into Switzerland.

The first leg of our escape was at night from Courcalles to Verdun by bicycle (about 30 miles) accompanied by a guide. When vehicles approached we all hid in the barrow pits or bushes. Our guide left us at a very large and beautiful home in Verdun. The home belonged to Count M. Pothier. He owned and operated a large factory before the German occupation and they allowed him to continue to manage it to produce war materials for the Germans. If they discovered he was a member of the French Underground it would surely result in torture and death for himself and the Countess.*

We were there for several days, during which time we were supplied with forged identification, work release and travel papers. We were to travel by rail from Verdun to Belfort (about 125 miles.) A young girl would guide us through the train terminals and along the city streets to our destination in Belfort. This was a daylight trip and we were to keep the guide in sight but never close enough to endanger her in case we were captured. I was given a newspaper to read (right side up, please) and Sylvia was to carry a handsaw in accordance with the work paper we had. We were to avoid conversation and just present tickets or papers on demand.

From Belfort, always escorted by one or more of the Underground, we traveled by wood-burning truck, bicycle and on foot, and finally through barbed wire fence into Switzerland.

The Swiss authorities released us into the custody of the American Attache. We were reunited with four more of our ten crew members (all enlisted personnel). As "Escapees" we remained in Switzerland for a year. As Allied Forces occupied France and neared the Swiss border, we again "escaped" to join the American Armed Forces.

*In July of 1978 I visited Verdun with my family. With much difficulty we were able to locate Mme. Pothier. She was living in near poverty in a small apartment. Would you believe that after all those years she did remember me and searched through her belongings to find a picture of me that she had kept.

They had been found out by the Germans. Her husband was tortured and put to death. She was tortured and then placed in a concentration camp. She was no longer a wealthy countess, but in Verdun there was a memorial to her family and a street had been named "Pothier."

I shall remember her always.

## THE MISSION BRIEFING
*by Elmer T. Lian, Lt. Col. USAF RET*

As a United States First Lieutenant in the Army Air Corp I was assigned as a pilot on a B-17 bombardment aircraft. I 1944 I was stationed in Mendlesham, England and assigne to the 34th Heavy Bombardment Group. Our mission was t fly strategic bombing missions over Germany and Germa occupied countries in Europe.

At about ten o'clock in the evening of the 26th of Septem ber 1944, an Army Air Corps soldier entered our "Nisse hut" and stated that our crew would fly a mission in th morning...

Our target would be the railroad yards at Ludwigshaver Germany, a large transportation center about 75 miles south west of Frankfurt. At this particular hour in history thi railway yard was busy sending hundreds of carloads c military supplies into the center of France. This was a bi factor in holding up General Patton's (U.S. Army) driv towards the German border.

Following the general briefing, it would be about 1:30 i the morning, the crews would head off for specialized brie ing. The navigator would get his special maps and tim "Hack," the bombardier would get the information on drop ping the bombs, the "Mickey" operator would get his radi signal information, every man would secure his parachut and special gear. On this mission our crew would lead 1,50 ships over Europe requiring several special officers an some new combat equipment on our new plane.

## THE MISSION

Little did I know when the wheels left the ground that i would be nine years before I would land again in Englanc Nor did I know that I had begun my last combat mission i World War II. And I was totally unaware that I was leavin a way of life and was starting on an adventure that woul include some of the saddest as well as the happiest moment of my life.

Your mind will always be working, thinking, wonder ing, fearing the unknown. You think of the mathematica changes which prove that you don't have an even chance o coming out of this war alive. Your reasoning goes along thi line; that five percent of the ships are lost on each mission an you are required to fly 25 missions. This is over 100 percen loss in the total number of missions. You think to yoursel that this is mission number 16—a fairly respectable numbe and in any group of flyers would stand up well in an conversation. But then again, this may be the last one.

As we headed into France at our altitude of about 27,50 feet, the contrails were forming behind all the ships. Thes contrails were a curse. If you could not see the planes from th ground you could see the white contrails, I would think t myself that if I were a gunner on the ground I would just ain and fire at the head of the contrails and, consequently, shoo the plane down. As an added problem, we were the firs plane, in literally an ocean of planes, and we were leadin 1,500 other bombers to a particular target area. If the gunne wanted anyone in particular, it would be the lead plane.

At 9:42 the tension had reached a high point. The pilo and co-pilot were now flying the ship in direct command o the radar officer who would drop the bomb. Directly ahead of our lead ship, I saw four large flak explosions. I knew they were out to get us and I said a quick prayer. Just then I heard a large explosion that sounded like a person had slammed a door in an empty house. I knew they had hit us. I glanced to the right and left and all four engines were turning. Then I noticed number two engine was smoking and in the center of the wing there was a large hole about a foot across. Then I saw that flames were streaming from the engine and the hole in the wing. This was it, the plane was on fire. In minutes it would explode into a ball of flame. I instinctively reached up and opened the cover on the bright red emergency bell button and pushed it repeatedly, about six times. Everyone knew what the problem was and what to do. It was all over; every man for himself.

## THE PARACHUTE JUMP

During my years in the Army Air Corps this was a moment I was not prepared for. No one had told me that a time would come when only one question would be presented to me—live or die.

I was now falling through the clouds. It was misting, wet and cold. I looked at my legs and noted that one of my shoes had come off in the sudden jolt of the parachute opening. My gloves were gone.

I was drifting down in an agricultural community. I noted several small towns and, to my surprise, I now noted some small cars that seemed to be moving in such a manner that they were boxing me in. I started swaying, but with a little trial and error with the parachute straps, I was able to halt the swinging motions. I thought to myself that I had been taught more how to stop the swinging of a parachute than I had on what to do as a prisoner of war. I could see Germans running towards my landing area.

As I lay crumpled on the ground and after I pulled the chute aside and saw I was surrounded by a group of perhaps 50 men, women, and young children. Several of the men had pistols or rifles pointed at me. Some were waving sticks, and some of the men and children had rocks in their hand. I attempted to stand but my legs gave way because my ankles hurt. For a moment I looked for mercy and understanding but those days were over, I did not know what they were thinking as the men waved the rifles menacingly at me, and indicated to me to fold up my chute. I must carry it and get moving. I was again on my feet despite the pain. By now several of the boys had thrown a few small stones, but the men held them back. I searched in vain for some soldiers because I felt maybe they would be more considerate. At this time I feared for my life. I didn't know where I would go or what they would do. They had no idea who I was, however, I assumed they knew I was an American flyer. I later learned that had been a false assumption. The place I landed was a cultivated garden about a block from a small village.

## GERMAN POW INTERROGATION CENTER

The place is known as Oberusel, Germany, and is known to every prisoner of war as a testing area for both the mind and body. I noted that I was in a situation reminiscent of any Army Induction Center. Men were lined up at various desks

and tables. I was first lined up to have my picture taken by a 35mm camera mounted on a high tripod. The operation reminded me of new recruits being taken into the Air Corps. Then I filled out a card giving my name, rank, and serial number. Then I went to a table where a Red Cross worker was seated. She had me fill out a card and advised me that as soon as I was identified this card would be sent to the Geneva Red Cross and my family would be notified that I was a prisoner of war.

It could be said at this time that one of the most meaningful items of my prison life was initiated by my writing a little card to my wife, Edwina. The card was simple but I knew there was no one in the world that would write a card like that and if she should receive it, she would know it was me and me alone.

I was about to undergo intensive psychological questioning... The major, in an offhand manner, asked my name, rank, and serial number, which I gave him. He then asked my duty aboard the plane. I informed him that as a soldier I was under a moral obligation to my country to give only my name, rank, and serial number.

He informed me in a pleasant tone that I was correct but that warfare had changed since that clause was written into the prisoner of war rules set forth in the 1929 Geneva Convention. He said it was impossible to establish who I was just because I was found in a parachute. He said thousands of foreigners were attempting to escape Germany by all known means. He advised me that it was absolutely necessary that I answer the questions he asked, so that he could send me to my permanent prisoner of war camp. He also wanted to inform the Red Cross of my welfare. He added that if I did not answer the questions he would turn me over to the German Gestapo, which would not be good.

Little did I know that on the afternoon of the tenth day, I would have my final interrogation. The guard came and called me; by now my hair was just a mat of lice and bugs, my beard was long and itchy, my hands had turned kind of shiny from never being washed, my clothes had become part of me. I smelled bad, but I could not detect it, and if I did, so what.

## PRISONER OF WAR RULES
## GENEVA CONVENTION 1929

The principal clauses governing care and treatment are outlined below:

1. A captured prisoner of war need give only the following information: name, rank, and serial number.

2. The food, clothing and shelter shall be equivalent to the Army of which he is a prisoner.

3. Medical and spiritual care shall also be equivalent.

4. Prisoners cannot be held in an active combat area.

5. Noncommissioned officers need not do manual work but may be used for supervising enlisted men at work.

6. Officers need not work.

7. Pay shall be equal to pay of men of equal rank in enemy Army.

8. Living quarters shall be plainly marked POW.

9. Transportation vehicles shall be plainly marked POW.

10. Solitary confinement will be considered punishment. A prisoner cannot be in solitary more than 30 days or 30

consecutive days for one offense. Three days must elapse before further solitary can be given.

11. Prisoners will salute men of equal or higher rank.

12. International Red Cross officers will act as inspecting officers.

13. Any questions arising from this treaty will be settled by the International Red Cross in Geneva.

*NOTE: The Germans did not consider rank in the Russian army and as far as the Germans were concerned, the Russian POWs were all "buck privates."*

Just before we boarded the train, to take about 100 of us prisoners from Frankfurt to Barth, Germany, the German guard and interpreter made the above speech to our group just prior to boarding the train.

The speech epitomizes much of the German thinking and the behavior expected from the prisoners of war.

It was the introductory speech to the POW life I lived.

## GERMAN GUARD SPEECH

Gentlemen you are prisoners of war, and will be treated in accordance with the Geneva Convention. You are to depart here and go to\_\_\_. The German officer is your commander, and he has orders from the station commander to deliver you dead or alive. He will fulfill his orders to the utmost.

You must remember you will be walking through the streets of towns and cities of which the majority of the people have lost most of their earthly possessions, lost most of their future, lost friends and loved ones, these people are very bitter against you men of American Air Corps, more so than anyone else in the world. They are easily aroused; therefore do nothing to incite them, the guards have orders to protect you. Do not sing, laugh, talk loud, or do anything that might arouse the civilian population.

According to the Geneva Convention you must be warned before you can be shot in your attempt to escape. You are hereby warned and therefore no further warning need be given. You are soldiers and prisoners of war. You will salute all German officers senior to you and regardless of the rank of the German guards. You will obey the orders given by them. I will be the interpreter and medical first-aid man on your trip, the senior allied officer will be \_\_\_, that is all.

## SUMMARY OF THE LIFE OF A PRISONER OF WAR

"The prison camp, in may ways, is a unique testing ground. The equality among the men is pitiless and unparalleled from any other type of organization. Family, culture, religion, wealth, education, and station in life makes no difference. A prisoner is on his own. Nobody will help or save him from the consequences of his behavior if it is deliberately wrong. He is forcibly placed among his peers in the most elemental circumstance and sparse environment. Here he is tested to see what he is made of. In the prison barracks he lives with prisoners of all types, coarse and refined, brutal and sensitive, rich and poor. He need not like them but they must all become a part of his life. Being a prisoner in a prisoner of war camp is a special way to participate in the affairs of ones time. The war was the common experience of my age and time. As history passes by, the least I can say is that I had a part of it."

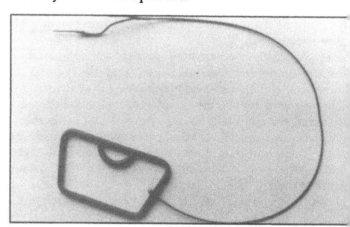

*World War II parachute D-ring, from chute of William B. Colgan, a Caterpillar member. (Courtesy of William B. Colgan)*

*A P-47 of the 86th Fighter Group, Europe, World War II. (Courtesy of William B. Colgan)*

# CATERPILLAR VETERANS

*Publisher's Note: All members of the Caterpillar Association and the Air Force Escape and Evasion Society were invited to write and submit biographies for inclusion in this publication. The following are from those who chose to participate. The biographies were printed as received with only minor editing. The publisher regrets it cannot accept responsibility for omissions or inaccuracies within the following biographies.*
*Photo: L to R, PFC Roger H. Hebner and Lt. Charles W. Haas 115th Observation Squadron, California National Guard. Taken 7 August 1932, posed by 0-17 airplane, identical to that from which they jumped on July 12, 1932.*

**LESTER E. ANDERSEN**, was born March 27, 1922 in Wadena County, MN. Joined the service in April of 1943 and had basic training in Fresno, CA, engineering for B-24 Bombers in Biloxi, MS. Gunnery school at Brownsville, TX. Crew training and assembly were at Westover Field, MA. Flew to Africa then Italy in August of 1944 with the 461st Bomb Gp., 765th Sqdn. as top turret gunner and flight engineer.

On his 11th mission was shot down over Munich, Germany Oct. 4, 1944 and escaped through Austria to Yugoslavia. Built dirt runway in Semic, Yugoslavia, radioed Bari, Italy and was picked up by DC-3.

After a short stay at Bari Hospital, returned to fly; plane was badly damaged after his 15th mission. Crash landed in Yugoslavia and after a short stay in the hospital returned to flying. On his 25th mission his badly damaged plane crashed at home base. After another short stay at the hospital, returned to complete 50 missions.

Returned to the U.S. in June 1945. Discharged and married Margaret Scolari in August of 1945. Became the parents of a girl and two boys. Presently have four grandchildren. Moved to Springfield, MA in 1950 and opened a construction company. Retired in 1985. His two sons presently manage the company.

**ROBERT A. ANDERSON**, M/Sgt. U.S.A.F. (Ret.), was born Jan. 23, 1916 in Northport, MI. Inducted in the U.S.A.F. May 15, 1941. Education: Northport High School, radio and electronics television schools and Allan Hancock College.

Training: aerial gunnery school, Buckingham Field, Ft. Meyers, FL; TV engineering, Los Angeles, CA; computer maintenance school, VAFB, CA.

Served with the following units: 865th G.S., Keesler Field, MS; 680th B.S., Tinian, Marianas; 439th Troop Carrier Wing, Selfridge Field, MI; 6th Maintenance Sqd., Walker AFB, NM; 1354th Video Production Sqd., Burbank, CA; 1605th AFB Wing, CSL TV, Lages Field, Azores; AFL-TV, Loring, AFB, ME; 576th Strategic Missile Sqd., Vandenberg AFB, CA; 3901st Strategic Missile Evaluation Sqd., VAFB, CA; and 4300th Support Sqd., VAFB, CA.

Achieved rank of master sergeant and served as a B-29 right gunner. Retired December 1, 1966.

Awards: Purple Heart, Commendation Medal with Oak Leaf Cluster, Asiatic Pacific Ribbon with two Battle Stars, American Theatre Ribbon, Good Conduct Medal with five Oak Leaf Clusters, World War II Victory with one Battle Star, American Defense Ribbon, Air Offensive Japan Ribbon, Eastern Mandates Ribbon, Gunners Wings and the Missile Badge.

Married Virginia in 1943. She died Feb. 28, 1984. They have two children and seven grandchildren.

Anderson worked for Federal Electric Corp. as senior technician, for Martin Company as quality control engineer and in the Corrective Action Branch of Spint Missile Program.

Today he is retired and lives in Punta Gorda, FL. He is the Florida State Commander of the Caterpillar Association, which held a Caterpillar reunion in Orlando, FL in January of 1988.

**EDWARD W. APPEL**, was born April 14, 1917 in Onamia, MN and was raised on a farm in Redfield, SD. Joined Army Air Corps July 29, 1940 as a private. Was promoted through the ranks to staff sergeant, then entered Aviation Cadets Aug. 4, 1942 and graduated May 20, 1943 as second lieutenant, class of 43E West Coast Training Center, Stockton Field, CA, where he had previously been crew chief and flight chief as enlisted man.

Trained as a pilot in B-25s, B-17s and B-24s. Flew B-24 Bomber to 8th A.F. via South America, Africa and on up to England. Flew combat against Germany in B-24s from February 1944 to Sept. 5, 1944 when shot down on 30th and last mission while bombing railroad yards at Karlsruhe. After about three months of evasion behind the lines, finally escaped first part of December of 1944.

Went back to England to 8th and volunteered for P-47 fighters with the 56th Fighter Group. Flew fighters escorting bombers, dive bombing and strafing until again being shot down strafing Muldorf Airfield, east of Munich on April 16, 1945. Again escaped after nearly being caught by German soldiers several times and made it back through

to our lines. Got back to Paris and the war w over.

Returned to the U.S. aboard an L.S. composed entirely of ex-POWs and evade in May of 1945. Left the service in Decemb of 1946, but stayed in the Reserves. Retired a lieutenant colonel.

Received two Distinguished Flyir Crosses, five Air Medals, four European Car paign Stars, the Purple Heart and sever other ribbons.

Married April 10, 1950 to home-tov girl, Crystal Crook, and had five children (; grown now) and 10 grandchildren. In civilia life was in fuel, oil and gas business an retired from that in April of 1984.

Still flying a rag wing Oironca which I restored in 1983.

**ROGER W. ARMSTRONG**, was bo Oct. 7, 1922 in Trumbull, NE. Enlisted Aviation Cadets when he was a student Augustana College in Sioux Falls, SD, an spent one semester in a college training d tachment at Creighton University, Omah NE. Class washed out as no schools we open when cadets reported to cadet center San Antonio. Bomber crews were needed f the 8th A.F. due to heavy attrition and the were made radio operators, armorers or eng neers.

Sent to Scott Field for six months of rad training. Three months followed in rad gunnery school at Yuam, AZ, then assigne to Lincoln Army Air Base with group f three months training as a crew.

Assigned to 8th A.F., 91st B.G. (H), 401 B.S. at Bassingbourn, England. Sent overse by ship due to lack of B-17s available Kearney, NE.

Shot down on 13th mission, Nov. 2, 194

on the way to Meresburg, Germany to the Luna synthetic oil refineries, by 88 flak shell that penetrated the plane behind the co-pilot and exited through the top of the plane, 30 minutes from target. Plane caught fire from ruptured hydraulic accumulator, severed oxygen, gas and hydraulic lines.

It was the world's largest air battle in history and 13 B-17s from the 91st B.G. were shot down over enemy territory, killing 41 men from the group (more than any group in 8th A.F. during its one-and-a-half years in the 8th A.F.).

Pilot John Askins received Silver Star for saving engineer and crew by putting out the fire on engineer's clothing and burning parachute. All bailed out, but co-pilot was killed by SS on the ground. Seven on the plane were awarded Purple Hearts. Armstrong bailed out of burning plane at approximately 25,000 ft., free fell 6,000 ft. before opening his chute. The plane blew up under him and lifted him about 30 to 40 feet in the chute. Captured by Germans then taken to a highway where two P-47s saw him and indicated they would strafe him. He pointed to the woods on both sides and indicated rifles were pointed at him. The leading P-47 wanted to attempt to land on the highway and pick him up, but as he flew back and forth, he indicated that he would be shot by German infantry.

Armstrong was incarcerated at Stalag Luft IV then moved to Stalag Luft I.

After the war he attended law school at Drake University and received an L.L.B. and a Doctor of Jurisprudence. He worked in claims management for 34 years with State Farm Insurance Company.

Married Gloria in June of 1946 and they have two children. Retired in April 1986 and now lives in Garden Grove, CA. Enjoys writing and traveling.

Discharged as technical sergeant. Decorations: Purple Heart, Air Medal with Oak Leaf Cluster, Good Conduct Medal, European Theatre Ribbon with two Battle Stars, American Theatre Ribbon, Victory Medal, and POW Medal.

## ROYCE F. AUSTIN,

went into the Army with the 75th Div. (as a medical aidman) to Ft. Leonard Wood, MO. Stayed with them until May 1943 then transferred to pilots training to Amarillo A.F. Base, TX. His whole barracks was transferred to Laredo, TX for gunnery training then on to phase training in Peterson Field, CO on to Omaha, NE to pick up their own B-24 to fly on to North Africa (Marakesh) then on to Spinazzola, Italy, where he started flying training missions.

On July 8, 1944, on their first mission (kept asking what this flak looked like) this veteran crew with 42 missions already under their belts said you'll see soon and they did. All hell broke loose. Austin counted about 10 B-24s

going down and approximately 60 chutes going down. They got hit over target (Vienna, Austria), at 18,000 ft. around noon on July 8, 1944. Tito's Partisans helped them evade and make contact to be picked up. After about three weeks they were picked up and sent to Bari, Italy.

Austin spent nearly a month in the hospital with a fractured leg. His original crew found out they were in the hospital and came to see them on their down days.

He asked to be sent back to his squadron. He told them he would stow away on a plane if they didn't put him back on crew. He had to go see Col. Babb and he said that he wasn't caught and they didn't know that he was over there being as he had escaped. Col. Babb finally let him go back to a crew upon his discharge from the hospital. He was placed on a crew who had lost their nose gunner on a mission. He flew 31 more missions with his new crew, until their misfortune on a direct hit over Blechhammer, Germany on Dec. 2, 1944 (oil refineries and marshalling yards).

After evading for several days he was taken prisoner and put in a stockade in Premiscel, Poland. He was questioned for six days as they did not believe he was "Americanski". They moved him to L'Wow, Poland and then to Poltava Russia.

They traveled to Siberia, Tehran, Iran, Tunis and back to Italy. He was back in his base on Christmas Day and was first in line and guest with Col. Price that day for Christmas dinner. He was grounded and shipped back to the States in February 1945. He went to Miami, FL and was sent to Washington, D.C. for interrogation for six days. Part of the parachute that he saved was turned into a wedding dress by his wife whom he married on March 18, 1945. Married to Imogene "Thomas" Austin for 46 years. Son: Dale A. Austin; daughters Dianne L. Shattuck and Janet L. Duhaime; six grandchildren, two girls and four boys and one great-grandson.

He stayed in the Air Force until Nov. 30, 1964 (22 years). He was selected as "U.S.A.F. Outstanding Airman" 1957.

From 1945-46 was recruiting officer in home town of Burlington, VT, transferred to Ft. Slocum, NY into the M.P.s Security and was teaching in the OSI Training School as small arms instructor. In 1949, he was transferred to the Armed Services Police Dept. Washington,

D.C. to form a cadre of all services. Retired Nov. 30, 1964 Larson A.F.B., Moses Lake, WA. When he got out of the service he was studying to become an undertaker but the man sold his business. He then applied for a job as chief of police in Maine and got it. Resigned to run for county sheriff but didn't make it (incumbent too strong).

He then went into business for himself as a general contractor, had trouble collecting checks so he quit. Then he went into the business of sand blasting and a bicycle shop and started selling monuments and engraving in cemeteries (it was said that this was a dying business).

Retired now and enjoys a leisure life of camping with a 40 ft. Holiday Rambler 7th Wheel and Ford crew cab.

## EDWIN D. AVARY,

Capt., was born Oct. 29, 1990 in Decatur, GA. Educated at Stanford University and Salzburg University in Austria/Germany.

Joined the U.S.A.F. June 15, 1932. Served with many different units at Randolph, Kelly, March, Hamilton, Crissy, France Fields and Cam Rahn Bay, Vietnam. Achieved rank of captain. Discharged May 1954.

Married and divorced twice. Has four children and seven grandchildren. Employed as chief pilot for Pan Am. Retired April 1, 1964. Now lives in Palo Alto, CA. Comment: "I'm proud to say that two of my sons are U.S.A.F. top gun fighter pilots."

## CHARLES H. BADER,

was born June 22, 1915 in Perkasie, PA. Joined the service on May 12, 1941. Achieved rank of first lieutenant (radar navigator).

Served with the R.O.T.C. at Drexel University, Aberdeen Proving Grounds, infantry training. Entered cadet training in 1945, Selman Field, LA. Received wings. Flew B-24 Bomber.

Qualified to be a Caterpillar when returning from 19th mission over Germany. After crossing the English Channel, a flare was shot to signal success. The gun was defective and the flare shot into the plane setting off a box of flares.

Captain Dawson, (pilot), told the men in the bomb bay to jump. Bader and three others jumped and were the only ones to make it. By the time the others were able to evacuate the plane, it was too low for the parachutes to open. Bader landed in a farmer's field and was returned to base with no serious injuries. Everyone thought he was dead and he had to collect all his belongings that were given away.

Units and Groups: 399th B.G., 605th B.S., Wendover Field, UT; 29th B.G., (May 1943), Squadron 6, Gouer Field, Boise, ID; 3rd Search Attack Unit (February 1944), 3rd Sqd., Langley Field, VA. In June 1944, flew to England on "Old 650" and became a mickey operator assigned to 392nd B.G., 8th A.F. Flew 21 missions before returning to States in June 1945 and was discharged.

When he returned to the United States, the young airman brought with him a piece of the parachute that had saved his life. His wife Eleanor made a christening dress, petticoat and cap for their first born, Cheryl Anne. Bader was very proud his life had been spared to give life to his child. When Cheryl Anne married she had two daughters who were also christened in the outfit. Said Bader, "Isn't life wonderful!"

Retired from Westinghouse Electric Company as a sales representative after 38 years. Earned a B.S. degree from Drexel University.

Decorations include the Air Medal, Distinguished Flying Cross with two Oak Leaf Clusters and four Battle Stars.

He and his wife, Eleanor have three children and nine grandchildren and have retired to Pocono Mountains in Pennsylvania.

**L.G. (BUD), BALLARD,** U.S.A.F, was born Nov. 23, 1914 in Boone County, MO. Enlisted in Army Air Corps Oct. 13, 1942 and was discharged Oct. 23, 1945 as technical sergeant.

Attended about all the major airplane mechanics schools in the States and was a qualified B-17 crew chief. Got shunted off into aerial gunnery school and became an aerial engineer gunner. Flew tailgun on B-17G Number 42-31831 named *Banshee.*

Went overseas with the 463rd BG (H) and became part of the 15th Air Force. Flew his first two missions out of Tunis, Tunisia, then moved to Celone Field, Foggia, Italy.

Was shot down on his seventh mission, April 6, 1944, northeast of Zagreb, Yugoslavia. Returned to his home base through the courtesy of Tito's Partisans

Spent the rest of his military career in a training command as an engineer instructor on B-17s at Ardmore, OK.

Retired as sales manager of Universal Foods Corp. Jan. 1, 1977. Has a son and daughter. He and his wife Irene live just one-half mile from where he was born.

**JOHN F. BARNACLE,** S/Sgt., was born March 30, 1919 in Winthrop, MA. Enlisted U.S. Army on Dec. 24, 1940; active duty Jan. 16, 1941, stationed at Camp Edwards, Cape Cod, MA with the 26th (Yankee) Div., 101st Field Arty. Later transferred to U.S. Army Air Corps,

Columbia Army Air Base, SC. Attended gunnery school at Myrtle Beach, SC and trained on B-25 medium bombers.

Volunteered for heavy bomber aircraft duty and was sent to advanced gunnery school at Salt Lake City, UT. In July 1943, while stationed at Clovis, NM, was assigned as an original member of the 450th BG Cadre and then was sent to Alamogordo, NM for group combat training.

After arriving in December of 1943, flew combat bombing missions with the 450th BG, 720th Bomb Sqdn., 15th Air Force, stationed at Manduria, Italy. Served as the armorer-gunner (ball turret gunner) on a B-24 Liberator Bomber.

On March 17, 1944, on 19th mission, Barnacle bailed out of his disabled aircraft while flying over Yugoslavia. Being declared MIA for approximately 30 days, he was subsequently rescued by Marshal Tito's Partisan soldiers who helped him to escape and evade capture.

He retired in 1980 from Redstone Arsenal, AL after 32 years of government service. He was employed as a budget analyst with the Dept. of the Army (Civil Service). He and his wife, Camilla, still reside in Huntsville, AL.

**FREDERICK E. BARNES,** born July 27, 1913 in Baltimore, MD. Inducted September 12, 1942 at Fort McHenry into U.S.A.F., then sent to Olmstead Air Depot in Middletown, PA, attached to 496th Sqd. Served as assistant crew chief on a C-39 cargo plane. Also played bass horn and bass violin in the post band and orchestra. After a year at Olsmtead Field, applied for gunnery school and was sent to Tyndale Field, FL. Graduated in December 1943. Assigned engineer gunner on B-17s and to Lt. Truman K. Eldridge's crew. Went through phases at Ardmore, OK.

Flew from Kearney, NE to Gremier Field, NH and on to Gander, Newfoundland and Nutts Corner, Ireland. Arrived in Ireland May 30, 1944 and was sent to Molesworth England, 303rd B.G. 359th B.S. Flew 24 missions.

Shot down at Meresburg, Germany on August 24, 1944. Was taken prisoner with left rib cage broken and sent to interrogation at Obereusel. Then sent to transport camp at Wetzlar and from there was sent to Stalag Luft IV in Kietheide, Germany.

Evacuated Feb. 6, 1945 and was on the

Black March that lasted 87 days on a diet of 70 calories "an ordeal that lives forever." Walke 700-800 miles. Faced rifle and pistol threat four times. Recovered by the 104th Wolve ines April 26, 1945. Weight had dropped from 170 lbs. to 108 lbs.

Crew survived due to pilot T.K. Eldridg who did not wait for bailout when he wa informed where they were on fire.

Achieved the rank of technical sergeant Discharged Oct. 20, 1945 at Fort Dix, NJ. Deco rations: Purple Heart, P.O.W. Medal, Good Conduct Medal, European Theater with three Battle Stars, Air Medal with three Oak Lea Clusters, and credit for D-Day, Northern France and Southern France.

Worked as an operating engineer and received a 50 year pin May 13, 1991. Retired and now active in various veteran's organiza tions. Married Clara (deceased) and has two children and one grandchild.

**ROBERT BARNUM (BARNEY),** LT COL., born Jan. 17, 1918 in Lake City, MI. Flying cadet, commissioned in October 1940 Kelly Field, 41-D. Served with the 64th B.S. 57th Pursuit Grp., 20th F.G.

Military locations: Flying Cadet, Spartar School of Aero., Muskogee, OK; Randolph Field; Kelly Field; Mitchell Field, NY; Bradley Field, CT; overseas via aircraft carrier Ranger Palestine; Egypt; Libya; Tunisia; ADTs Albu querque, NM; Corsica; Italy; (end of WWII) Drew AFB, FL; Dover AFB, Ft. Leavenworth, KS; 2nd Army Hq. Baltimore, MD; 11th A.F. Hq. Harrisburg, PA; 20th Fighter Grp., Shaw AFB; Maxwell AFB, AL; Hq. USAF; UNC FEC Hq. Tokyo; CINC PAC Hq. HI; B-47 combat crew course, McConnell AFB, KS; Savannah AFB, GA; Plattsburgh A.F.B., NY.

Planes flown: Funk 75; Aeronca Champ; Cessna ISO; Cessna 195; Citarbara; PT-19; PT-22; BT-14; BC-1; AT-6; L3; L4; L5; C-45;

C-47; UC-78; C-123; A-20; B-5; B-34; B-47; P-38; P-39; P-40; P-47; P-51; F-84.

Electrifying experiences: forced landing 100 miles southwest of Khartoum, Sudan, in fuzzie wuzzie land (P-40) July 1942; shot down strafing in Libya, landed in no-man's land (P-40) January 1943; hit cement pole in hazy conditions strafing convoy, Po Valley (P-47) January 1944; bailed out south of Bologna, Italy after severe electrical storm knocked out all gyro instruments (P-47) April 1945.

Awards: Silver Star; Distinguished Flying Cross with Oak Leaf Cluster; Air Medal with six Oak Leaf Clusters; Purple Heart; Distinguished Unit Citation with three Oak Leaf Clusters; Air Defense Service Medal; American Campaign Medal; European/ME Campaign Medal; WW II Victory Medal; National Defense Service Medal; and French Croix de Guerre.

Married Chris Bryant Dec. 14, 1947, and they have three children and one grandchild. Currently residing in Charlevoix, MI.

Earned M.B.A. at Gonzaga University in Spokane, WA 1967. Retired Oct. 31, 1963, Plattsburgh AFB. Keeps moving. Mows lawn, plays golf and shovels snow.

## GORDON H. BICKLEY, was born Sept. 26, 1921 in Detroit, MI. Joined U.S. Army Air Corps as Aviation Cadet Sept. 8, 1942.

Cadet training was at Kelly Field, San Antonio, TX and Stanford, TX. Assigned to A.A.F.G.S. Las Vegas, NV; then to CIS, Fort Meyers, FL. After completion was assigned to BG at Drew Field, Tampa, FL. Joined crew as ball turret gunner on B-17s, in February of 1944.

Arrived E.T.O. May 1, 1944. Assigned to 452 BG, 729 Sqdn., 8th Air Force. On Bickley's 18th mission, July 8th, plane was severely damaged by flak over Rouen, France. After bailing out, he evaded capture and was hidden by several French families until area was taken over by Canadian ground forces.

After returning to U.S. he was assigned to A.A.F.C.S.F.G., Laredo, TX, then to K.A.A.F. Kingman, AR as gunnery instructor. Discharged Sept. 3, 1945.

Married Donna Carswell June 16, 1943. They have four children and five grandchildren and reside in Redford, MI. Bickley re-

tired in December of 1983 after careers in industrial advertising, sales promotion and auto parts manufacturing.

## JOSEPH R. BILY (BOB), born Nov. 11, 1922 in Berwyn, IL. Joined the Army Air Corp July 1942. Achieved rank of 2nd Lt., navigator and was discharged Dec. 14, 1945. Earned a B.B.A. degree from Southern Methodist University in 1949.

Trained: Selman Field, Monroe, LA. Served: Polebrook, E. Anglia, England with 8th A.F., 351st B.G., 509th B.S., B-17s.

On Jan. 19, 1944, on a formation training mission prior to going to England, his B-17 was leading the formation back to base. The plane to the left was attempting to change position in the formation and came up underneath his. The other plane's top turret crushed the nose of his plane and Bily was knocked out of the destroyed nose section. Fortunately he had on a back-pack parachute.

The pilot, 1st Lt. Fred Heiser, returned the plane to base without the nose, but the other B-17 crashed, killing eight of the crew. One crew member bailed out. Awards: Air Medal with three Oak Leaf Clusters, Distinguished Flying Cross, European and American Theatre, and Victory Medals.

Married to Lila Marie and they have three daughters and three grandchildren.

Retired May 1, 1987 as president of actuarial and benefit consulting firm in Houston, TX.

Currently a part-time consultant residing in Arlington, TX. Enjoys travel and golf.

## DALE E. BOGGIE, Col., entered the U.S. Air Force in November 1950 after hostilities broke out in Korea. Because all quotas were filled for entry into pilot training, he enlisted as a buck private and went to basic training at Lackland AFB, TX.

Upon graduation from basic, he was assigned to aircraft maintenance school, Shephard ABF, TX. With two years of college and an aircraft mechanics license prior to entering the Air Force, he graduated with honors and was selected as an instructor.

Rapid promotion to staff sergeant and selection as a course supervisor in the jet engine mechanic school, Amarillo AFB, TX,

followed. Repeated attempts to enter pilot training were unsuccessful, so Boggie applied and was selected for officer candidate school. A "Distinguished Graduate," he was commissioned as a second lieutenant and again applied to pilot training. Instead he got more of the "exigencies of the service" routine and was sent to aircraft maintenance officer school then to Korea as an F-86 aircraft maintenance officer with the 8th Fighter Bomber Wing. He finally made it to pilot training when he returned from overseas and graduated at the head of his class in May 1956. He received a coveted assignment to the F-86 Fighter Gunnery at Williams AFB, AZ, proving that perseverance pays off.

Became a Caterpillar in May 1957. He was a first lieutenant flying at 40,000 ft. over New Mexico on a flight from El Paso, TX to McConnel AFB, KS, flying an F-86H Sabre jet on a beautiful sunny day. The instrument panel warned of fire in the engine compartment and black smoke was boiling out from underneath the fuselage near the trailing edge of the wing. There was no time for a mayday call. Boggie squeezed the ejection seat triggers, felt a terrible jolt and tumbled like "a rag doll." When he hit the desert ground he tumbled and rolled, driving his heels up through his buttocks and also took a nasty crack on the head (cracked his helmet). The wind started dragging him and the chute over rocks and cholla cactus plants. He was finally able to pull in enough shroud lines to gather the chute and sit on it. He had no broken bones but looked like a porcupine with all the cactus sticking in him with barbs like fishhooks. He was alone on the desert for three hours before being picked up by an elderly Mexican in an old pick up truck.

During his 28 year career in the Air Force, Boggie flew 26 different types and models of aircraft and logged over 4,300 hours, including 315 hours of combat time. He served as squadron and group commander and was on the Air Staff in the Pentagon. He retired as deputy base commander, Lowry AFB, CO.

Following retirement, he formed a corporation engaged in contract audio-visual services to the government.

He and his wife Sharon now reside on a small horse ranch near Parker, CO.

**MARCEL BOISVERT,** born Aug. 16, 1925 in Concord, NH. Entered the U.S. Army Air Corps in September 1943. Received gunnery training in Las Vegas, NV and combat crew training in a B-17 as a tail gunner at Avon Park, FL. Served with the 8th A.F., 381th B. G., 534th B. S. in Ridgewell, England.

February 14, 1945 on fourth mission to Brux, Czechoslovakia, his was a B-17 named "The Fox". Going to, over the target and on return flight, the plane was severely damaged by flak. Unable to keep up with the formation, were forced to drop back, losing sight of any planes. They flew alone through Germany with three of four engines feathered. Plane was hit by more flak and the crew was ordered to bail out. The fourth engine was on fire.

Boisvert left the plane through the waist hatch. He came down over a German anti-aircraft battery and they were waiting for him. Civilians in the area took his shoes and parachute. The first four of the crew to land were captured. The remaining five crewmen landed over the lines.

Eight members of the crew are alive today.

Discharged in October 1945. Re-enlisted in October 1946.

Discharged again in August 1949. Education: 1950-52, attended the University of New Hampshire; 1952-56, Tufts University School of Dental Medicine; 1964-66, Boston University School of Graduate Dentistry, specialty Orthodontics; 1974, Diplomate American Board of Orthodontics. Still in practice. No thought of retirement. Married Memorial Day 1955 to Barbara Bower. They have three children and two grandchildren. Currently resides in Reading, MA. Enjoys antique cars, photography and learning H.A.M. radio.

**WILLIAM P. BOOTH,** born in May 1924 in Carsondale, PA. Joined U.S.N.R. on April 1, 1943. Attained the rank of technical sergeant.

Served with the 446th B.G., 105th B.S., U.S. Navy VR-24 in P.T. Lyautey during Korea, assignment S.P., M.A.A., Dog Catcher. Flew B-24s.

Graduated Scranton T.H.S. 1941, Penn-

sylvania Upstater until February 1942 - Philadelphia and Levittowner.

Basic training in Miami Beach; radio, Scott, gunnery, Tyndall, crew assignment, Westover, Mitchell, Goose Bay, Nutts Corner, Stone, Bungay.

Qualified to be a Caterpillar on 29th mission, Sept. 26, 1944, Hamm, 24,500 ft. Hit by flak. Captured and taken to Luft IV.

On Black Death March. Twelve member crew: Mullery, Kallstrom, Jones, Watt, Lengel, Gold, Cochrane, Bush, Cotton, Haugen, Booth. (Charron didn't make mission. Rest POWs.)

Discharged in October 1945. Awards: European Theatre Operation, Air Medal with four Oak Leaf Clusters, POW, and Caterpillar.

Retired in 1977 from a U.S. government position.

**ROBERT G. BOOZER,** born Aug. 10, 1922 in Europra, MS. Joined service Jan. 2, 1940. Served with 155th Inf. and 15th A.F. in North Africa, Italy and Southern France. Flew most all planes, including Combat B-24.

Discharged July 15, 1945, rank of lieutenant.

Married Oct. 29, 1940 to Dee. They have two children and two grandchildren.

Civilian employment: owner and operator of advertising firm.

Member of American Legion. Resides in Franklin, NC.

**EUGENE B. BOWARD,** was born March 2, 1921 in Leitersburg, MD. Joined military Sept. 3, 1942. Photo-gunner on daylight and night-intruder missions in A-20 light bomber, 86th Sqdn., 47th BG., 12th A.F., from bases in Italy, Corsica, and southern France.

Shot down on 51st mission over Apennine Mountains of northern Italy at 1:30 a.m. Nov. 12, 1944 while participating in night intruder (single aircraft) mission to destroy German pontoon bridge across the Po River. Parachuted, joined Italian Partisans, and evaded southward through the U.S. 5th Army front lines on Nov. 20, 1944 in company with two other crew members (Lt. Dowdell and S/Sgt. Schultz), Partisans, escaped P.O.W.s, other Allied evadees and six or eight German soldier defectors. Discharged Oct. 25, 1945.

Civilian engineer with U.S. Army. Retired in 1986. Lives with wife, Sue, in Springfield, VA. Son: 1st Lt. Gary D., U.S. Army.

**HERBERT A. BRADLEY,** was born Nov. 13, 1920 in Cleveland, OH. Became an aviation cadet in July 1942 in U.S. Army Air Corps.

Served with 8th Air Force. Locations: Kirtland Field, Albuquerque, Biggs Field, El Paso, TX, Midland AAB, Midland, TX, San Angelo AAB, San Angelo, TX, Gowen Field, Boise, ID.

Bradley's 453rd B.G., 2nd Div. (B-24s) led the entire 8th Air Force to Brunswig, Germany on May 8, 1944. They survived attacks by hundreds of Me-109s and FW-190s and made it back to England when forced to jump from battle damaged aircraft in Watton, England.

Farmer met him with a pitchfork. When Bradley finally convinced him he was not a German the farmer remarked, "I didn't know you boys were practicing parachute jumping today." The young airman convinced him they were not. Awards: Distinguished Flying Cross, Air Medal with three Oak Leaf Clusters. Applied for Caterpillar membership in England.

Married Mary Cynthia Norris. Daughter's name is Janice Renee Bradley DeMoss. Grandson is James Bradley DeMoss. Granddaughter is Amanda Marie DeMoss. Stepson is Jon DeMoss.

Active in Oklahoma Wing of Confederate Air Force. Semi-retired branch manager for collegiate cap and gown company located in Oklahoma City, OK, selling choir robes, pulpit and judicial gowns.

## ARDEN ANDY BRENDEN, was born

Jan. 8, 1922 in Starbuck, MN. Enlisted Oct. 14, 1942. Was in Air Force 407th Bomb Sqdn. A.A.F. Went to mechanic school in Amarillo, TX, gunnery school in New York and engineer school in Boeing Factory, Seattle, WA. Achieved rank of staff sergeant.

Was shot down on first mission over Chalones, France. Landed in field near a wooded area. There was a church nearby where a wedding was being performed. Landing interrupted the wedding as everyone stopped to watch the parachute landing. Was helped by French families until he reached French Underground. Underground helped him get over the Pyrenees Mountains. Entered Lerida, Spain May 15, 1944. Discharged Oct. 14, 1944.

Was awarded the Purple Heart, Good Conduct, E.T.O. two Battle Stars.

Married Pauline on Oct. 20, 1945 at Alexandria, MN. Have four sons and four grandchildren.

Was field rep. for Marquette Cement for nine years. Owned and operated Culligan Soft Water business at Canton, SD for 18 years. Retired after 18 years as administrative officer for Farmers Home Adm. Currently an A.A.R.P. instructor in 55/Alive. Live in Mesa, AZ six months during winter and in Huron, SD the other six months.

## FREDERICK G. BRODERSEN, was born

Jan. 15, 1923 in Baraboo, WI. Educated at Hamline University. Joined the U.S. Navy June 3, 1942. Served at various locations with various units. Had many memorable experiences. Flew 12,000 hours, mostly singles. Still flying FH-1 1949, first jet. Flew just about all singles.

Discharged Aug. 1, 1963 with rank of lieutenant commander. Married six times. Has four children and one grandchild. Owns

Brodersen Business Forms and lives in Tampa, FL.

## LOUISE BOWDEN BROWN, was born

May 16, 1916 in Over Brook, PA. Learned to fly in a small field in South Jersey. Joined W.A.S.P. in March 1943. Trained for six months in Sweetwater, TX and then was assigned to the 3rd Ferrying Group in Romulus, MI. Delivered small planes before receiving a white card from instrument school in St. Joseph, MO. Sent to fighter school in Brownsville, TX. Had excellent training and flew the AT-6 from the back seat with instructor in the front seat. Flew DC 3, five single engine fighters and light planes (51, 47, 39, 63, 40).

Picked up a Mustang in Texas and stayed that night in Greensboro, NC. The next morning the group took off for Newark, NJ. The aircraft's engine sputtered and they turned to head back. The engine continued to sputter, ran smoothly then quit. The plane was on fire and Bowden landed in mud in a corn field. A Navy plane was circling overhead.

Discharged Dec. 20, 1944. Married Sam Brown (deceased) on May 15, 1987.

Employed as a medical technologist, U.P. Philadelphia. Also missionary in India and Nepal. Retired in July 1986. Now resides in Hickman, KY. Enjoys gardening, golf, travel and church.

## MARVIN W. BROWN, was born Oct. 8,

1925 in McClean County, KY. Served in World War II, inducted Nov. 2, 1943, achieved rank of staff sergeant and Korean War, inducted April 26, 1951, achieved rank of 1st lieutenant, 303rd B.G., 8th A.F., 440th F.S. Flew B-17, F-80, F-84 and F-86.

Shot down over Holland in World War II. Was POW for seven months. During Korean Conflict, bailed out of F-86.

Medals: Air Medal, Prisoner of War Medal and Purple Heart.

Discharged Nov. 25, 1945 and again March 1, 1956 with rank of 1st lieutenant, U.S.A.F. (retired).

Married Marian Walker and has five children and four grandchildren.

## LLOYD J. BURNS, was born April 28, 1923

in Haywood City, NC. Joined the U.S. Air Corps Dec. 14, 1942.

Was stationed in England, A.F. Station,

118. Served with 8th A.F., 392th B.G. Achieved the rank of staff sergeant and was discharged Oct. 10, 1945.

Awards: Distinguished Unit Citation, Air Medal with Oak Leaf Cluster, Purple Heart and POW Medal.

Most memorable experience: Being on a bombing mission to the "most valuable single target" in Germany's two-engine manufacturing complex. The target at Gotha, Germany was placed on the strategic priority list and the 392nd was assigned to destroy it. They did.

The Gotha aircraft industrial complex was virtually all but destroyed by "a superior bombing attack with a 97% precision bombing accuracy". The 392nd earned a Distinguished Unit Citation for the mission, but the toll was high. Seven of their planes were lost.

After the bomb run fighters viciously attacked for over two hours. Burns' plane suffered flak and fighter damage. It lost two engines on the same side and caught on fire in the bomb bay. The crew was forced to bail out.

Burns retired Aug. 10, 1979 from U.S.A.F., Mac/Air Weather Service. Currently is employed by G.E. Government Services as manager of data processing.

He is single and lives in Canton, NC.

## THORNTON LEE CARLOUGH,

P.O.W., M.I.A., graduated as a second lieutenant pilot, class 43J at George Field, IL Nov. 3, 1943. Was in 1st Replacement Crew, 460th BG, Spinazzola.

Returning from Bucharest, April 15, 1944, crossed Danube with two engines out and eased B-24 into a 180 to avoid mountains and bailed. Landed inside Bulgarian border with two other crew members. Held by Bulgarians. Together with gunner escaped into Yugoslavia. (Other member was shipped to Stalag 111).

Seized by a small band of Chetnick guerrillas. Traveling over goat paths, high narrow ledges with precipitous drops, and sheer rock faces, interrupted by a skirmish between the Chetniks and Ustachi, reunited with the seven crew members who landed in Yugoslavia. Furnished with sten guns and Chetnik guides, the nine covered over 200 miles indescribable terrain avoiding many German patrols. After four weeks, connected with a B-17 crew downed in January. All trekked to reach British Underground Unit Force 399 CMF, British

Intelligence, ordered by London to diverge from Mihailovich. On orders of Gen. Armstrong (British) Lts. Al Romans (pilot of B-17) and Carlough were furnished with horses and instructed to select the best of five suggested landing sites. The strip at Pranjne was chosen, some plum trees removed. On May 29, hundreds of Chetnik, led by Mihailovich himself, surrounded the field to protect the people who were deserting them, from approaching Germans. Smoldering kerosene lamps lined the strip and an R.A.F. ship made an incredible blacked out landing to evacuate them all to Bari.

After seven weeks in occupied territory, returned to combat flying June 3rd. Received Cluster on Purple Heart July 15th on fourth trip over Ploesti. Returned stateside in October. Instrument school Bryan, TX, instrument instructor Napier Field, AL until separated in August of 1945.

Resides in East Hartford, CT.

**JOHN PHILIP M. CARLSON,** was born April 16, 1915 in Shickley, NE. Entered Aviation Cadet Program in August 1942 at Santa Ana, CA and completed training at Hondo, TX. Assigned as a navigator to the 460th BG. Operational training at Clovis, NM; Kearns, UT; and Savannah, GA.

Flew combat missions from Italy in B-24s and was shot down May 18, 1944, in Yugoslavia returning from a bombing mission over Ploesti, Rumania. Evaded capture with aid of the Chetniks and returned to base August 10, 1944.

Reassigned as an aviation psychologist with the Air Force Redistribution Command at Atlantic City, NJ. When separated from active duty in 1945, held the rank of captain.

Awards included the Distinguished Flying Cross, Air Medal with Oak Leak Cluster and Distinguished Unit Citation. Retired from the Air Force Reserve in the rank of lieutenant colonel.

Married Maryjo Suverkrup Oct. 14, 1950 and they reside in Fairfax County, VA.

Now retired from a career in the federal government and the practice of law. Remains active on a part time basis in government liaison matters.

**BENJAMIN B. CASSIDAY, JR.,** was born July 25, 1922 in Honolulu, HI. Became a cadet in the U.S.M.A. July 1, 1940, commissioned June 1, 1943.

Was a member of the 87th F.S., 79th Fighter Group; 27th F.G.; 81st F.G., 36th F.W.; Air Grp.II (USN).

Turkey. Flew P-40, P-47, P-38, P-51, F9F (Navy), F4D, F-104 and F-01.

Became a Caterpillar in April 1943 when he bailed out of stricken AT-6 in New York state.

Most memorable experience: First mis-

sion on May 31, 1944 over northern Italy when his flight of four P-47s was jumped by 12 Me-109s. After many twists and gyrations he shot down one Me-109, but became separated from flight during the dog fight.

The airman's plane was badly shot up, (especially the radio). He finally landed in friendly territory, then returned to base in Corsica.

Later flew 116 missions in Italy, France and Germany. Also served as a Navy exchange pilot in Korean War and flew 43 missions in F-9F Panther off the carrier "Philippine Sea".

Cassiday was advisor to the Turkish Air Force and also deputy commandant at the Air Force Academy and commandant of Air Force R.O.T.C. program. Served at the U.S. Military Academy and National War College.

Discharged Nov. 30, 1972 with rank of brigadier general.

Awarded the Air Force Distinguished Service Medal, Distinguished Flying Cross, Legion of Merit with three Oak Leaf Clusters, Air Medal with eight Oak Leaf Clusters and Service Medal with one Battle Star.

Married Suzanne, (second), in 1983. He has two children and together they have nine grandchildren. Resides in Honolulu, HI.

**JAMES C. CATER**, was born July 4, 1922 in Chicago, IL. Served in the U.S.A.F. Joined May 4, 1942.

Served in England in 1944; Guam from 1948 to 1949; Okinawa in 1950-51. Units: 91st B.G. in E.T.O. and 19th B.G. in Korean War.

Shot down in France in a B-17 on April 28, 1944 on 21st mission over Ajord, France. Bailout at 15,000 ft. Flew 55 missions on B-29 in Korea.

Achieved the rank of major and was discharged in 1956.

Married Frances R. Cater. Has three children and seven grandchildren. Received education at the University of Southern California.

Is vice president of a commuter airline and president of a municipal bus company. Resides in Poughkeepsie, NY.

**JOHN C. CATHERINE, SR.,** was born Oct. 7, 1923 in Morrisdale, PA. Attended high school there and later the Pennsylvania State University, earning B.A. degree. Became member of the U.S. Army Air Corps on Oct. 19, 1942.

Training: basic at Keesler Field; radio at Chicago; gunnery at Tyndall Field; 21st Anti sub Sqd. at Gulfport; Ephrata, WA, to form 483rd B.G.; trained as group at McDill. Served in 817th B.S., 483rd B.G. (H).

Most memorable experiences: flying into Macon, GA with B-17 (a chemical warfare field with Piper Cubs); being shot down and captured at the Isle of Capri; liberation day (May 2, 1945) and arrival at home.

Discharged Oct. 11, 1945 with rank of technical sergeant.

Married Josephine Menger Feb. 23, 1951 and has five children and four grandchildren. Civilian employment: public welfare casework, unemployment compensation clerk, rehabilitation counselor, aging representative, area director of public welfare. Retired Sept. 30, 1983. Resides in Jersey Shore, PA.

**LEE H. CHALIFOUR,** enlisted at age 16 in AFS, 1944. Initially attached to British 14th Army with Air Evac unit, later transferred to U.S. Army Air Force as a B-24 tail gunner, 14th A.F., CBI.

Involuntarily exited aircraft returning from bomb mission on Jan. 17, 1945 over the jungles of Burma, due to hostile action. Over seas assignments in India, Burma, China, Newfoundland, Panama, Hawaii, Japan, South Vietnam, and Germany.

Awarded 24 ribbons and decorations, including Bronze Star, Air Medal, Purple Heart, Meritorious Service Medal with Oak Leaf Cluster, Air Force Commendation Medal with OLC, Vietnamese Cross of Gallantry with Palm, three medals from Great Britain, China War Medal, Chinese A.F. gunners wings.

Participated in five major campaigns. Retired USAF as a medical administration superintendent, rank of chief master sergeant, with 30 years of service in the U.S. Air Force.

Wife Mary Lee retired as a lieutenant colonel, USAF, NC after 20 years service. The Chalifours have six children and five grandchildren and now reside in Port Charlotte, FL.

Lee Chalifour is a member of the 14th Air Force Association, Hump Pilots Association, China-Burma-India Veterans Association, C.A.F., Burma Star Association, D.A.V. and American Legion. He has served terms as local commander and state commander, national junior vice commander-SE, and national historian, CBIVA; as vice president-East, 1st American Br., Burma Star Association the past 11 years.

**JOSEPH J. CHAPAS,** was born Nov. 3, 1916 in Aurora, IL. Joined U.S.A.F. on April 15, 1941. Served with the 58th Mat'l Sqd., McChord Field, WA, 15th A.F., 301st B.G.; 32nd B.S., Italy December 1943 to February 1944. Flew twin engine Beech B-17G.

The young navigator qualified to be a Caterpillar the first time in May 1943 flying from Mather Field to March Field. A ram in the morning forced bailout over Fontana in heavy fog.

The second bailout occurred Feb. 25, 1943 near Regensburg, Germany due to German fighter planes. Six of ten crew members were lost.

Decorations: Air Medal with five Oak Leaf Clusters, POW Medal, Purple Heart, American Campaign, Air Reserve, E.T.O. Campaign with three Battle Stars, WWII Victory, American Defense, Res. Longevity Ribbon, and Presidential Unit Citation with three Oak Leaf Clusters.

Chapas is married and has one son, Micheal and one grandchild, Jevin. Micheal was a Cobra pilot during the Vietnam War.

Chapas retired from the Reserves in 1980 with the rank of major. He enjoys golf, bowling and cross country skiing.

**DONALD H. CHAPLIN,** was born Nov. 13, 1922 in Winchester, MA and graduated from high school in Melrose, MA in 1940. Joined the service on Nov. 11, 1942. Served

in the U.S.A., Europe, South America, Okinawa, Thailand, and Vietnam. Served in U.S.A.A.F. and U.S.A.F. Served in WWII with the 27th Fighter B.G. and later A.D.C., ATC, TAC and U.S.A.F.E. Achieved the rank of lieutenant colonel.

On Sept. 12, 1944, Chaplin was flying number four position in a ground reconnaissance flight of four P-47D Thunderbolts into the Belfort Gap region of eastern France, chasing a German armored division. It was his 12th (and last) WWII combat mission.

The P-47D-28 took a few rounds of AA fire which set the aircraft ablaze. He turned back toward friendly territory descending rapidly to get more speed and distance. The cockpit walls disappeared in flames so he bailed out at well over 300mph and close to 150 feet AGL (according to a witness). The "beautiful nylon parachute canopy" snapped open with a tremendous shock just as his right foot hit a soft farm field furrow. Chaplin was badly burned and broke a few bones and later commented that it was "an exciting day".

Chaplin was discharged in December 1945 and again in September 1969. Later employed with Federal Aviation Administration in flight inspection, and was a pilot for 16 years. Retired from FAA in August 1986.

He and his wife Harriet Thyberg Chaplin have three children and three grandchildren and reside in Nashua, NH.

**WALTER E. CHAPMAN, SR.,** was born Feb. 16, 1921 in Lexington, KY. Entered service in July 1942, U.S.A.A.F. Aviation Cadet. Graduated class 43-10, Midland, TX as a bombardier, rank second lieutenant. Discharged in July 1945, with rank of first lieutenant.

Served overseas with 484th BG, 826th BS, 15th A.F. Credited with 24 combat missions on B-24 type aircraft from May 5, 1944 to June 13, 1944, when shot down by German Me109. Parachuted out of aircraft over Italy and with assistance of Italian and Yugoslavian partisans evaded capture until August of 1944. Two crewmen killed in action, four taken prisoner of war and Chapman, along with three other members of the crew evaded capture with help of Tito's Partisans.

Retired industrial engineer with 32 years of service with International Harvester Co.

**RALPH N. CHRISTENSEN,** was born Feb. 11, 1921 in Philadelphia, PA. After graduating from high school, went to Johns Hopkins University for two years and studied engineering. Joined the USAAF March 26, 1942. Locations and stations: Maxwell Field, AL; Dorr Field, Arcadia, FL; Selman Field, Monroe, LA; Laredo AAF, Laredo, TX; Lowry Field, Denver, CO; Mountain Home Army Airfield, MT. Home, ID.

Served with the 825th B.S., 484th B.G., 49th Wing, 15th A.F. in Cerignola, Italy September 1944 to May 1945 and with the 1010th AAF Base Unit, Atlantic City, NJ August-September 1945. Achieved the rank of staff sergeant. Discharged Oct. 2, 1945.

Married Jan. 5, 1943 and has five children and three grandchildren.

Christensen worked for Reid-Avery Company in Dundalk, MD, a welding rod manufacturer for 28 years, as head of inspection and quality control. Later he worked for Bethlem Steel Company, Sparrows Point Shipyard in Sparrows Point, MD as a welder for 12 years. He retired Feb. 11, 1983. Now resides in Baltimore, MD.

**FORREST S. CLARK,** was born July 6, 1921. Joined the service Oct. 5, 1942. Served with the 44th B.G., 8th A.F. in Shiphdam, Norfolk, England.

Flew with B-24 combat crew. Awarded Air Medal and Purple Heart. Discharged Oct. 9, 1945 with rank of technical sergeant.

On a return from a bombing mission to Oslo-Kjeller, Norway, his formation was attacked by approximately 15 enemy fighters over the North Sea. They lagged behind and were the target of most of the fighters. The fighters came from all directions and many attacked the tail position. Forrest was the tail gunner and was busy firing the twin .50 caliber machine guns mounted in the tail turret. An enemy shell went through the turret and struck one of the gunners who was bringing ammo back to Forrest's post, wounding him in the head and arm.

After losing considerable altitude, the plane went into a cloud bank and the enemy fighters lost them. The plane was headed directly into the North Sea but at the last minute managed to level off. The command was given to jump into the frigid water but fortunately they managed to limp back over the coast. The crew was ordered to bail out as the bomber was so badly damaged and had lost the landing gear.

The date was Nov. 18, 1943 and Clark was the radio-operator-gunner on the crew of Lt. R.C. Griffith, 44th B.G., 67th B.S., 8th A.F. Two engines were out and the landing gear was damaged but the pilot crash-landed the plane, saving the life of the wounded gunner.

Married Ruth in June 1948. They have two children and two grandchildren. Retired as a news writer/journalist in July 1986. He remains a WWII aviation buff and now resides in Kissimmee, FL.

**ALTON W. COCKRELL (Al),** Major, USAF (Ret.), was born Dec. 16, 1921 in Sanford, NC. Enlisted NCNG in February 1940. Mobilized Sept. 16, 1940, transferred to Army Air Corps in March 1943.

Cadet Class 44-A. Flight Instructor Waco AFB, P-47 Fighter CBI 33rd Fighter Gp., 59th Fighter Sqdn., 80th Fighter Gp., 89th Fighter Sqdn., 86 combat missions. Emergency jump on 82nd mission. Evadee. Rescued by infantry and 89th commander Major Rankin.

Awards: Air Medal with two Oak Leaf Clusters, DFC and Presidential Unit Citation.

Discharged Dec. 10, 1945. Alumni of the Class of 1950. Recalled to active duty in June 1951. Retired as a major from the USAF in 1966.

Aircraft checked out in: PT-19, BT-13, T-6, L-5, U3A, L-20, C-47, P-40, RA-24, P-47, T-33, F-86, RC-121, (3000 hours) Civilian Cessna 150, 172, 177, 182. Commercial Pilot still flying in 1990.

Worked for 20 years with the state of North Carolina after military retirement. Office Manager. Civilian retired in October 1, 1985. He is a twelve gallon blood donor. Volunteers for many city, count, and civic club projects. He is married to Corrine, they have two sons Tony and David and two grandchildren. They reside in Sanford, N.C. He is a member of the Air Force Escape and Evasion Society, Daedalian and Caterpillar Association.

**WILLIAM B. COLGAN,** Col., U.S.A.F. (Ret.), was born Dec. 1, 1920 in Quitman, GA. Enlisted Dec. 12, 1941 from Waycross, GA, Aviation Cadets, Class 43-C, pilot wings, March 1943.

WWII service in Mediterranean and Europe, 79th and 86 Fighter Groups, flying 208 combat missions, P-40 and P-47 aircraft, as pilot, flight leader and squadron commander.

Colgan was leading a mission from the base near Nancy, France on Feb. 28, 1945 to attack rail targets in Germany. Near Zweibrucken, an enemy train was spotted loaded with armor and his group started dive-bombing the target. Pulling up from the attack, his plane took a direct hit and the P-47 instantly started down, losing altitude quickly.

The airman struggled to get free of the plane and once would have been thrown out if his leg had not been caught in the cockpit. He was able, however, to get the aircraft back somewhat under control and limped toward friendlier territory. There was a large hole in the left wing and several smaller ones in the left side of the cockpit. There was also a direct hit in the tail. The instrument panel was in shambles and the pilot's leg was injured.

Once over his own field, he decided he must bail out. He tried to climb out but his leg was hung in the cockpit. He pull back to loosen it and successfully lunged out again. The tail of the plane flashed by inches from his face. Parachuting down, Colgan realized that the plane was almost directly under him. He hit the ground in freshly-plowed French farmland, barely missing the fiercely burning hole in the ground that had been his P-47.

He later served in the Korean War, 136th and 58th Fighter Wings, flying a combat tour in F-84s as squadron commander. Also served and flew combat in Vietnam. Other key positions: chief of test flying, Eglin A.F.B., FL and director of operational requirements (future airplanes, weapons), TAC, Langley A.F.B., VA. Retired 1972 from position, commander 326th Air Division, Hawaii.

Awards include: Silver Star, Legion of Merit with OLC, Distinguished Flying Cross with three OLC, Purple Heart, Air Medal with 14 OLC, French Croix de Guerre (SS), numerous service and campaign medals, also the wings of an Army parachutist. Colgan is the author of the highly-acclaimed book "World War II Fighter-Bomber Pilot".

He is married to Anita Lamae Allen and they have two children and two grandchildren. They reside in Shalimar, FL.

**JAMES B. COLLIER,** was born September 25, 1920 in Ironton, OH. Joined the service March 17, 1941.

Training command as flier bombardier

cadet and was a B-26 instructor pilot fror 1941 through 1945 with 42nd B.S., 504th B.G 313th Wing, B-29s.

Planes flown include: PT-19, BT-13, A1 C, BC-1, B-10, AT-11, B-26, B-24, and B-29.

Jumped from a B-29 after a strike o Osoha, Japan on June 7, 1945. Was in wate about eight hours before the Navy picke him up. Five of the crew were lost, six wer rescued. Collier was the airplane com mander.

Discharged in early 1946 with the ran of captain. Was awarded Air Medal and th Purple Heart.

Married Bette and has two children an four grandchildren. Semi-retired lawyer.

**WILLIAM COOK, JR.,** Col. U.S.A.F. (Ret. was born Sept. 14, 1943 in Greenwood, MS Drafted U.S. Army 1943, retired regular U.S Air Force 1973.

Flew unarmed T-6 Texans in combat i Korea as forward air controller (FAC). Flew 69 types of aircraft (B-17, B-29, B-25, B-47(150 hours), C-47 (4000 hours), F8F, F-51, P-4( F2H Banshee, T-33, transports, tankers, etc.

Served in operations and command as signments in SAC, Hq. U.S.A.F., and MAC Awarded Legion of Merit, Air Medals, Meri torious Service Medal, Commendation Med als of Joint Services, Army and Air Force Japanese Occupation, Korean Service wit three Battle Stars, and many others.

Parachuted from a B-25 Billy Mitchel bomber in 1947. Aircraft exploded in flight

Married to Katherine, daughter of Ad miral L.A. Moebus, U.S.N. (deceased), a na val aviator. Five children. Cook resides i Mary Esther, FL.

**JOHN F. COONEY,** was born Nov. 1! 1923 in Brooklyn, NY. Enlisted in A Com pany, 106th Infantry Regiment, 27th Div NYNG, on Oct. 14, 1940. The unit was feder alized in January 1941 and became A Bat. 186th Field Artillery, 6th Corps, A.U.S. Joine the U.S.A.A.F. Oct. 14, 1942. Completed aeria gunnery school and aircraft armorers schoo and washed out Aviation Cadet Training i September 1943. Completed air crew train ing December 1943.

Arrived in E.T.O. March 1944. Assigne to 332nd B.S., 94th B.G. (H), 8th A.F., sta

tioned at Rougham Aerodrome, Bury St. Edmonds in East Anglia, England. Was a tail gunner on a B-17 G with the rank of sergeant.

On Sunday, April 23, 1944, while flying a practice mission the aircraft caught fire in the cockpit area, the bailout bell sounded as the aircraft flew erratically. The waist door would not release and in the few seconds before the aircraft stalled off on its left wing, only the radio operator (S/Sgt. Harry P. Sonnett) and Cooney managed to bail out of the tail hatch door. The rest of the crew were killed in the ensuing fiery crash. The incident happened one day after Cooney had flown his first mission to Hamm, Germany. Conjecture was that the aircraft had sustained hidden flak damage that caused the oxygen supply to the upper turret to mix with lubricant and catch fire.

Cooney flew six combat missions before being grounded with a stomach ulcer in June 1944. He transferred to the infantry at the time of the Bulge in December 1944 and served with the 3rd Army, and 354th Inf. Regt., 89th Div. to the end of the war. He saw action at the end of the Battle of the Bulge, the Hurtgen Forest, the Rhine at St. Goar, and central Germany. Attached to Combat Team 4, liberated first concentration camp at Ohrdruff, Germany, the scene of absolute horror. Wounded while on a task force clearing Lossnitz, Germany in May 1945.

Awarded American Defense Medal, American Campaign Medal, Good Conduct Medal, European/African/Middle East Campaign Medal with six Battle Stars, Bronze Star, Purple Heart, Air Medal, WWII Victory Medal, Army of Occupation Medal with Germany Clasp, The Medal of Liberated France, Presidential Unit Citation with OLC, the Combat Infantry Badge and the Aerial Gunners Badge.

Discharged at Fort Dix, NJ on Oct. 14, 1945. Returned to school, earned B.S., employed by the city of New York for 27 years until retirement in November 1978, as civil engineer/land surveyor. Relocated from New York to Massachusetts and later returned to work, employed by Dept. of Army at Fort Devens, MA from June 1982 to June 1990. Retired for second time from position in construction management.

Married in January 1951 to Margaret (nee Deitz); two children, a son, John Vincent Cooney, (pathologist), and a daughter, Anne-Marie Cooney, (attorney). Has two grandchildren, John Francis and Teresa Ann Cooney.

Has membership in: V.F.W., D.A.V., American Legion, 94th Bomb Group Memorial Assoc., 89th Div. Assoc., Caterpillar Assoc.

Memorable experiences: first mission - witnessing in disbelief the disappearance of

a B-17 caught by a flak burst in a ball of fire and black smoke over Hamm, Germany; loss of crew and events of April 23, 1944; crossing the Rhine River March 23, 1945 in the pitch black early morning assault and being dumped into the river (managed to reach enemy shore when enemy artillery and machine gun fire opened up when they reached liberating Ohrdruff concentration camp and seeing the atrocities committed by the SS.

**PETER COTELLESSE,** Col., U.S.A.F. (Ret.), was born April 15, 1922 in Trenton, MI. Entered Aviation Cadet Training September 1943 in Detroit, MI. Commissioned and rated a pilot at Pampa A.A.F., TX April 15, 1945. Service in Air Transport Command, Far Eastern Air Forces, Air University and Air Training Command.

Assignments included chief, Wing Operations Center, 3rd Bomb Wing, Johnson A.B., Japan; chief, Operations and Training, A.F. Section, U.S. Military Supply Mission, India; professor aerospace science, Auburn University, AL; director of curriculums, Air University and director, military and specialist training, DCS technical training, air training command. Combat tour, Vietnam, 1967-1968.

Awards include Legion of Merit with OLC and Distinguished Flying Cross. Retired August 31, 1974 after 30 years of service.

Life member of the Retired Officers Assoc. (TROA) and Order of Daedalians. Served as national adjutant, Order of Daedalians, June 1978-June 1989.

Married to Jeanne Vecchiato, Detroit, MI. Has four children and four grandchildren.

**SILAS M. CRASE,** D.M.D., M.P.H., A.B., F.A.C.D.; Col. (ret.), was born Jan. 30, 1925. As a 97th BG, 414th Sqdn., 15th A.F. S/Sgt. ball turret gunner, flew 11 combat missions from Foggia, Italy. Bailed out Aug. 27, 1944 after bombing Blechhammer oil refineries, Germany. Evaded out of Germany into Czechoslovakia. Joined and worked with Slovakia Partisans in Zilina/Vella Byctcia area.

With "Russian Liberation", the Russians urged that he be shipped to Moscow for

repatriation. Evaded the "Russian Allies" after about three weeks. With a deaf-mute passport traveled across war wrecked Czechoslovakia and checked in with the American ambassador in Prague. Then transferred to the U.S. Army in Pilsen June 15, 1945.

Worked with Kentucky and Ohio state health departments. Retired in January of 1985 from U.S. Army Dental Corps with a total of 30 years.

Married Anna Marie Auton. Children: Deborah Ann and Michael and five grandchildren.

Retired in Fort Pierce, FL.

**WALTER S. CROWELL, JR.,** Capt., was born March 24, 1924 in Melrose Park, PA. Volunteered Feb. 1, 1943. Served with 15th A.F. (Italy), 456th B.G., 745th B.S.

Shot down over Blechammer, Germany, Oct. 13, 1944, flying 30th mission. Wounded in left eye by flak burst that splintered the B-24 bomber's windshield. Narrowly escaped execution by enemy firing squad before imprisonment at Stalag Luft III. One of three survivors of B-24 crew who embarked from U.S. in April 1944.

Awards: Purple Heart, Air Medal with five OLC, POW Medal.

Separated from service Jan. 6, 1946.

Married Virginia and lives in Littleton, MA. They have three children and six grandchildren. A graduate of the University of Pennsylvania, Crowell is a software consultant.

Most memorable mission: August 15, 1944, bombing invasion beaches, southern France, and bearing witness to horizon-to-horizon armada of U.S. Navy ships carrying troops and supplies to open a new front.

**JASPER C. CROWLEY,** Greenbank, WV. Merchant seaman from 1937 to 1941. Entered U.S.A.F. April 29, 1942. Overseas July 4, 1944. Assigned to 95th B.G. (B-17). Stationed at Herham, England. Flew 33 combat missions. His first 25 with Lt. G. B. McVay. Most of that crew was lost on a training mission Nov. 6, 1944. Flew next seven missions with Lt. C.H. O'Reilly as bombardier. Shot down by Me-109 on 33rd mission to Hamburg. Plane blew up. Part of the crew perished. Crowley bailed out at 25,000 ft.

just before plane explosion. Evaded capture for 24 hours. Taken prisoner January 1, 1945, with wounded leg and frost bitten feet.

Held in solitary for a week then sent to Wetzlar, Nurmburg and finally Moosburg. Freed by Gen. Patton's 3rd Army.

Awarded: Crew and Bombardier Wings, Presidential Unit Citation with two OLC, Purple Heart, Air Medal with five OLC, American Theatre Medal, European Theatre Medal with one Silver Star for five major battles - Normandy, Northern France, Rhineland, Ardennes, Air Offensive Europe, Good Conduct Medal, Victory Medal, POW Medal, Polish Home Army Cross and the Hellenic Air Force Commemorative Medal from the government of Greece.

Discharged Sept. 30, 1945. Retired after 31 years working for the state of West Virginia as a state park superintendent.

**J. J. CUFF, JR.** was born Dec. 29, 1949 in Patuxent River, MD. Joined the U.S.M.C. June 4, 1975. Earned B.S. at the U.S. Naval Academy and his M.S. at Salve Regina College.

Served with VMAQ-Z, MAWTS-1, USCINCPAC/ABNCP. Achieved the rank of major and is still in active duty.

Cuff ejected April 28, 1983 during night FCLP; when aircraft quit flying (EA-6B). Everyone survived. His second ejection came four months later to the day at almost the same time. This time the aircraft (also a EA-6B) erupted in a fireball and destroyed all controls. Everyone ejected safely.

Married Terry M. Cuff and resides in Honolulu, HI.

**EARL O. CULLUM,** Col, (Ret.), was born in 1913 in San Antonio, TX and was raised in Dallas. Attended North Texas Agricultural College and Texas A&M College. LLB degree, Blackstone School of Law; graduate of U.S. Army Military Police School, Army Command and General Staff College, F.B.I. Academy.

Commissioned U.S. Army Reserve 1937, entered active duty March 1941, training new 208th Military Police Company, Camp Bowie, Texas, Instructor at U.S. Army Military Police School. Overseas September 1943 on Stilwell staff in China-Burma-India Theatre. Later was commanding officer of 159th

Military Police Battalion. Promoted to lieutenant colonel and earned BSM and China War Memorial Medal. Non-combat highlight was personally capturing deserter-murderer-escapee-fugitive who became the only American executed in C.B.I. theatre in WWII. Reverted to Reserve status in December 1946. Promoted to colonel in 1957 and retired in 1967 with 30 years Reserve service.

Spent 30 years as F.B.I. special agent, retired 1977. Career highlight was personally capturing a "Top Ten" bank robber fugitive wanted coast-to-coast, who received a 30-year sentence. National commander of C.B.I. Veterans Assoc., Texas department commander of Military Order of World Wars, president of 500-member Rotary Club of Dallas, director of Greater Dallas Crime Commission. Son, Col. Richard O. Cullum, 1961 U.S.M.A. graduate, served in Vietnam, now retired. Son, Kenneth H. Cullum, Purdue R.O.T.C. honor graduate, served in Vietnam, now president of Bristol Steel Company in Virginia. Daughter Kathryn C. Lee resides in San Antonio. Cullum has eight grandchildren and five great-grandchildren.

**CLAYTON C. DAVID,** Lt. Col. (Ret.), was born July 19, 1919 in Topeka, KS. Joined Army Dec. 1, 1941 and eventually served in U.S. Army Air Corps, U.S.A.F.

Took pilot training at Gulf Coast training command bases; combat training, Pyote, TX and Dyersburg, TN. Flew B-17, C-47, P-51, P-47, P-63, trainers. Released from active duty Dec. 11, 1945. Retired from Reserves July 19, 1979.

Flew with 358th Sqdn., 303rd BG, 8th A.F., Molesworth, England, WWII. Shot down as co-pilot on mission's return over Holland May 25, 1944.

Returned to U.S. and worked with Intel-

ligence; then assigned to Great Falls, MT and Long Beach, CA, (for ferrying planes).

Married to Lenora M. (Scotty) Scott Fe 11, 1945 and has two children, Lynn A. and James S. David and one grandchild, Jonatha David.

Was district field superintendent of P Inc. and associate professor at West Virgin Northern Community College. Retired fo last time Aug. 1, 1985. Currently resides Hannibal, MO.

**WILLIAM W. DRISKO (BILL)**, wa born March 15, 1925 in Bartlesville, O Joined Aviation Cadets March 1943. Receiv pilot's wings Jan. 7, 1944. P-51 training Bartow, FL. Arrived in England at Bodn Airdrome on June 5, 1944. Assigned to 487 F.S., 352nd F.G. Flew escort and strafir missions until June 25th. Got in front of a Me-109 that shot off the right wing so baile out. Had some difficulty getting out of plan with only half of a wing. Picked up by Frenc Underground and hid from Germans un September. Liberated in Paris.

Assigned in States to various fighte units flying P-51, P-40, L-5, B-25, T-6 and B' 13s.

Relieved from active duty in Octobe 1945. Graduated from Oklahoma State Un versity in January 1949. Worked for Phillip Petroleum Company in sales and was re called to Korean Conflict at Great Falls, M' flying C-54. Sent to Korea December 195 flying C-47, T-6 and C-54. Relieved from active duty in December 1952. Returned t Phillips Petroleum Company and remaine for 21 years. Worked as special account supervisor for Echrich Meats Company fo eight years. Retired in 1987. Stayed in A.' Reserves 20 years. Retired as major in Marc 1970. Retired for pay March 15, 1985.

Married and has three sons and tw grandchildren.

**WILLIAM C. DuBOSE,** was born Marc 27, 1924 in San Francisco, CA. Joined Arm Air Force in November 1942 and was sta tioned at many bases and locations. Was P 38 pilot. Achieved rank of first lieutenant.

Received Air Medal, Purple Heart an E.T.O. Ribbon.

Married to Deanna. Children: David

Desiree, Steve and Ken. Planned to retire in November 1991 after 36 years with 3M in sales and marketing of aerospace products.

**DANNY R. DUFF,** was born Sept. 20, 1924 in Lexington, KY. Joined the service in March 1943. Served with the 386th B.G., 9th A.F. (WWII) and Air Weather Service, 5th A.F. (Korea).

Stations: Barksdale A.F.B., LA; Taegu, Seoul, Nagaya, Japan, Misawa, Japan (Korea), England, France, Belgium (WWII). Flew B-26, A-26, B-25, C-47. Discharged in September 1952 with rank of captain.

Married Mary T. in 1947 and they have five children and ten grandchildren.

Earned a degree in metallurgy and worked for Reynolds Metals and I.B.M. Retired in May 1987.

Currently likes to travel, volunteers for A.A.R.P., and enjoys planes and computers. Resides in Lexington, KY.

**E.F.W. EISEMANN (BILL)** Col., (Ret.), was born Aug. 30, 1923 in Stuttgart, Germany. Entered U.S. Army Jan. 28, 1943. Served in the 8th A.F., 446th B.G., 704th B.S., 5th A.F., 6147th Tactical

Control Grp. (Korea), 5th A.F., 460th Tactical Reconn. Wing (Vietnam).

Bailed out April 10, 1945 while on a mission over Germany. Aircraft (B-24J) was seriously damaged by flak immediately after dropping bombs. For a while it seemed they would have to leave the plane over Germany, but the pilot, with "courage skill and the grace of God", managed to nurse it back to England. The damage was of such nature, however, that the crew was forced to bail out.

Awards: Legion of Merit, Distinguished Flying Cross with ten OLC, Bronze Star, Air Medal with three OLC, Meritorious Service, Air Force Commendation, Good Conduct/ Army, Presidential Unit Citation with OLC, Air Force Outstanding Unit/"V" and OLC, Air Force Organizational Excellence Award, American Campaign, American Defense Service, E.T.O. with three Battle Stars, Vietnam Service with four Battle Stars, Armed Forces Reserve, Korean Presidential Unit Citation and United Nations Service Medal.

Planes flown: B-24 (navigator), T-6, B-25, C-45, C-47, U-3A, C-54, B-29 and C-130.

Discharged A.A.F. May 10, 1946. Recalled for Korea Aug. 15, 1950. Retired from A.F. May 1, 1975.

Earned M.S. degree in management from Columbia University and retired from TRW Feb. 1, 1989.

**PAUL FEDELCHAK**, was born June 22, 1917 in Brownsville, PA. May 1937 Army Air Corps technical school, Chanute Field, IL. Units: 3rd Balloon Sqd., 91st Observation Sqd., Flight "F" 1st Mapping Grp., 2nd Photo Grp., 15th Photo Sqd., 28th Reconn. Tech. Sqd., Headquarters F.E.A.F., and 8th Reconn. Tech. Sqd.

Military schools: Chanute Field, IL, squadron officers course, AL, Eastman Kodak, Ansco Corp., (ret.). First parachute jump out of O-46 in 1940. Forced landing in Dawson, NW Territory Dec. 15, 1941, F-2 twin-engine Beachcraft.

Returned from aerial mapping of Alaska, Dec. 8, 1941 and ordered to patrol NW Pacific Ocean, daily, 50 miles out. There was a massive Pacific storm and there was no one in authority to cancel the flight. Three were on board and flying conditions were unstable with the plane flying through ocean waves. High pressure oxygen bottles and toolbox broke loose and the contents were flying around the cabin. The mission was aborted and two men bailed out.

The emergency door jammed and Fedelchak shot off the hinges with a 45 side-arm and jumped. He blacked out but revived, parachute invisible, flying vertically over Olympic Forest. He dragged through the tree tops, clothes shredded and shoes gone and crashed into a tree top. Climbed down and jumped to the ground then stumbled blindly through the forest. "Captured" (rescued) 25 miles inland at Huptulip by Army troops and civilians who were patrolling the forest and shooting at "suspected foreign paratroop invader".

Advisor to French in Hanoi and Den Phen Phou, 1952. Discharged Aug. 30, 1958 with rank of major. Awards: Legion of Merit, Bronze Star with OLC, Commendation Medal and Presidential Unit Citation.

Married Betty Aug. 12, 1942. They have four children and four grandchildren and live in Livingston, MT. Fedelchak retired

Aug. 30, 1958 and now enjoys golf, fishing and traveling.

**LOUIS FEINGOLD,** was born July 11, 1919 in Brooklyn, NY. Drafted January 1941 into the Infantry, transferred to the Air Corps, June 1942. Completed navigation training, Mather Field, Sacramento, CA, June 1943.

Flew 20 combat missions in B-17s in E.T.O. Shot down Dec. 30, 1943, 60 miles north of Paris. Was immediately aided by the French Underground who hid him and helped him escape along their Shelbourne Line. Arrived by boat in England, through Operation Bonaparte, on Feb. 27, 1944. Returned to U.S. and served as navigational instructor until discharge as a lieutenant.

Following discharge was self-employed as a garment contractor in New York where he remained until December 1990. He is now semi-retired. Married Leah in 1954; they have four children.

**EDWARD J. FERRARA, SR.,** was born Oct. 17, 1924 in Springfield, MA. Joined the Army Air Corps in April 1943. Was technical sergeant radio operator/gunner, crew 38, 467th B.G., 789th B.S., E.T.O. (England), 8th A.F. Was at Scott Field (radio) Yuma (gunnery), Westover (crew training) and Langley (sub patrol).

Flew 35 missions, May 1944-April 1945 in B-24s.

Forced to bail out during Dec. 29, 1944 mission. Plane was damaged on take-off during fog (Battle of Bulge and fog all over Europe). Four planes were lost attempting take-off to drop frags (anti-personnel) on German troops/positions - Ardennes (Prum).

Awards: Air Medal with four OLC, E.T.O./African/Middle East Theatre and Good Conduct Medal. Discharged Sept. 5, 1945, Plattsburg Convalescent Hospital in New York.

Worked as a manufacturer's representative. Has five children and six grandchildren.

**WAYNE S. FISCHER,** was born Nov. 25, 1940 in Milwaukee, WI. Enlisted in the U.S. Army Nov. 26, 1957, Commissioned May 13, 1965. Served with: 1st, 7th, and 10th Special

Forces Groups (Airborne), 101st Airborne Div., 1st Aviation Brig., 7th Inf. Div., 8th Inf., Div. (LRRP) and other separate units worldwide.

Aircraft flown: CH-47A,B,C, CH-34, OH-58, CH-54A, OH-23, O-1, UH-1A,B,C,D,H, CH-19, AH-1G, OH-6, OH-13, C-47 and numerous civilian aircraft types. Code name: "Bushmaster Six" and others.

At the time of his emergency bailout, Fischer was a 15-year-old Civil Air Patrol cadet. He had attended a CAP cadet exchange in Canada the previous week. He missed his return flight connection Aug. 13, 1955 in Toronto, Canada and had been lucky enough to hitch a ride in a Royal Canadian Air Force Beaver aircraft. The pilot had agreed to drop him off at the Detroit, MI airport to catch a commercial flight home to Milwaukee, WI.

At about 9,000 ft. over southwestern Ontario, the engine caught fire and the pilot elected to bail out. Fischer was understandably reluctant to jump but the pilot "convinced" him.

The young man did not know how to steer the chute and landed like "a sack of over-ripe tomatoes" in a farm field directly in front of a tractor that almost ran him over. That was his first and last emergency jump. He later enlisted in the U.S. Army and became a paratrooper and now has over 800 parachute jumps.

Fischer became a rated helicopter pilot after completing three ground combat tours in Vietnam. He was shot down twice and force landed nine times and commented, "Army helicopter pilots are not issued chutes".

Awards: Silver Star, Distinguished Flying Cross, Bronze Star "V" with two OLC, Air Medal with 16 OLC, Meritorious Service Medal, Army Commendation Medal with OLC, Army of Occupation Medal-Berlin, Good Conduct Medal-3rd Award, Vietnamese Cross of Gallantry, and other campaign and unit citations including the Army Broken Wing Award.

Retired from the Army Jan. 31, 1979 with rank of major.

Married to Hilde and has five children and eight grandchildren. Resides in Colgate, WI. Employment: On-Hand Industrial Ad-

hesives sales manager. Earned B.S. and M.B.A. degrees.

**M. HEARTY FITCHKO**, was born Oct. 12, 1921 in Ligonier Township, PA. Enlisted in Aviation Cadet Training Program April 13, 1942. Commissioned and rated a pilot in May 1943.

Served in England, Spain, Morocco, Japan, Alaska, California, Florida, Kansas and Ohio. Retired with rank of colonel Feb. 28, 1966.

Fitchko set a speed record Maine-England in a B-47 in 1952 and won an SAC bombing competition in a B-47 in 1957. He had a forced bailout in a B-17 Oct. 1, 1944 and another in a T-33 jet Dec. 11, 1952.

Awards: Distinguished Flying Cross with OLC, Air Medal with three OLC, Joint Services Commendation plus 14 others.

Married Polly Graham March 31, 1951 and they have two children (son and daughter) and four grandchildren. The Fitchkos own Ligonier Sales and Service, Retail Case/IH tractors and farm equipment and Cub Cadet lawn and garden equipment. Says son will be taking over very soon. Resides in Ligonier, PA.

**LEONARD W. FRAME,** Lt. Col., U.S.A.F. (Ret.), was born Oct. 27, 1917 in Selma, CA. Graduated California Polytechnic University in 1938, San Luis Obispo, CA with a major in dairy husbandry.

Entered flying cadets in March 1941, class 41-H, primary and basic flying, Cal Aero Academy, Ontario, California. Advance flying, Stockton, CA, graduated Oct. 31, 1941.

Assigned to 70th Pursuit Sqd., Hamilton Field, CA. Orders to sail for Philippines Dec. 8, 1941 were aborted. Sailed for Fiji Islands Jan. 12, 1942.

On a routine training mission at Fiji in P-39, as Frame flattened out at the bottom a loop, there was a bang and the airpla began to vibrate. Soon he felt heat around right foot and called his wing man and to him he was "leaving this thing". He roll out the door and saw the airplane flami out the right side just before it hit, going with a "big whoosh".

Frame landed on a small hill and s there until his wing man came flying by a he stood up to wave at him. He rolled up t chute and walked to the top of the little hi He ran into a G.I. who had been working a road crew. He saw the airman and start running. It took them about a half an hour walk back the distance the G.I. had run about 10 minutes.

One of the squadron pilots came out a jeep to pick him up and commented th "everyone you crack up (airplanes) mak you a better pilot." Frame felt that he didr want to improve his piloting skills that wa

He went on to Guadalcanal where flew 79 combat missions in P-39s and w credited with destroying 1 1/2 Zeroes ar 1/2 Betty Bomber.

Awards: Purple Heart, Air Medal wi six OLC, Distinguished Flying Cross. Pr moted to first lieutenant in March 1942, ca tain in March 1943 and major in Octob 1944. Released from active duty in Septen ber 1945. Active in Air Force Reserve un 1963. Promoted to lieutenant colonel in Ju 1955. Retired from Air Force Reserve in O tober 1977.

Purchased a farm in Fresno, CA in Marc 1946. Operated as a dairy until 1976. No producing almonds. Married to Cynth Claybaugh Jan. 7, 1942 and has two childre and three grandchildren.

**KENNETH GARWOOD,** was born No 29, 1921 in Saranac Lake, NY. Joined th service May 15, 1942. Was a radio gunne with the 8th A.F., 96th B.G., 339th B.S. flyin B-17s.

Shot down over Magdeburg, German June 20, 1944. Hit by FW-190 and Me-10 German fighters. Both inboard engines wei burning and the airman bailed out at 20,00 ft. and delayed chute opening for about 15,00 ft. Spent 10-and-a-half months as POW an participated in the 600-mile Death March.

Awards: Air Medal with OLC, E.T.O. Medal with two Battle Stars and POW Medal. Married to Gwen for over 45 years and has six children and 13 grandchildren (all well). Semi-retired jeweler. Discharged Oct. 3, 1945 with rank of technical sergeant.

## WALTER G. GRAF,

was born Feb. 17, 1920 in Philadelphia, PA. Graduated Franklin and Marshall College, Lancaster PA. Received his wings at Moore Field, Mission, TX in the class of 44F and P-47 Thunderbolt. Operational training at Abilene, TX. Assigned to the 27th FG, 523rd Fighter Sqdn., 12th Tactical Air Command at Pontedera, Italy in the winter of 1944-45. The 27th was transferred to St. Dizier, France, in February of 1945 to supply tactical air support to the Seventh Army on the drive for the Rhine.

On March 20, 1945, "Coalbox" Red Flight of four Thunderbolts looking for targets of opportunity crossed Speyerdorf, the main Luftwaffe field in the German Palatinate at tree-top level. All four Jugs caught heavy flak and Graf bailed out at minimum altitude when his plane caught fire and exploded. After hiding in the mountains the rest of the day, he walked in a westerly direction at night hoping to make it to France. On the second night he was captured by Wehrmacht artillerymen and taken to German headquarters.

After an unsuccessful attempt at interrogation he was sent en route to a prison camp somewhere deep in Germany. The car in which he and his captors were traveling was caught by American artillery fire and destroyed. Proceeding on foot for two more days, Graf, who could speak German, became acquainted with his guards. On the second night he lured one of the guards aside, clobbered him and took off. He was successful in getting through to the 100th Infantry Div. of the American Seventh Army which had just cracked the Siegfried Line in the area. After a hot meal and some medical attention for a wounded leg, he was returned to base by an Infantry L5. From there he was returned to the States on Project "R".

Today, Graf is retired. He lives in Fort Washington with his wife Irene. He has one daughter, two sons and five grandchildren.

## BILL HALL,

was born October 1942 and inducted into USAF Feb. 17, 1943. Assigned to 381st BG, 1st Air Div., 8th AF, P-51 Scouting Force.

Joined the service October 1942/Feb. 17, 1943. Served in the U.S.A.A.F., 381st B.G. (B-17), 1st Air Div. (8th A.F.), Scouting Force (P-51).

Served at many various military locations and achieved the rank in active duty of first lieutenant and Reserve rank of lieutenant colonel. Earned the Air Medal with six OLC.

On April 12, 1945, impatient to see Joe DiMaggio and his touring baseball team play at Bassing Bourn, Lt. Hall split-essed from 26,000 feet. Coming out of the second split-ess the Mustang attained compressibility and could not be induced to abandon its vertical attitude. One wing decided to go down by itself causing the remainder of the plane to go into an outside spin. Sensing that he had lost control of the situation, Hall decided to bail out. He unsnapped the safety belt, neglected to dump the canopy, discard his oxygen mask or disconnect the radio. Centrifugal force prevailed which caused Hall to become known as the pilot who was strained through plexiglass. During his overnight stay in the base hospital, he was informed by an orderly of the death of President Roosevelt.

Discharged from active duty with the rank of first lieutenant and from Reserves with the rank of lieutenant colonel. He received the Air Medal with six Oak Leaf Clusters. He is an engineer at Boeing, Everett, WA.

## LOUIS L. HALTOM,

was born June 14, 1919 in Nacogdoches, TX. Joined the military Sept. 8, 1940 and achieved the rank of lieutenant colonel, U.S.A.F.

Education: College, 1938-40 Stephen F. Austin University (Business Admin.), Nacogdoches, TX; Texas Christian University 1951 (Public Law), Fort Worth, TX; Landman's Assn., 1961 (Minerals Laws), Wichita, KS. Military schools: (1940-41) Air Corps training command, primary basic and advance single engine pilot training, Sebring, FL; (1943) B-29, first pilot check-out training, AMC, Midwestern Procurement Command, Wichita, KS; (1945) single engine jet pilot and maintenance training, San Bernardino, CA, U.S. Army; (1947) food service supervision course, San Francisco, CA; (1951) U.S.A.F. Air University, Craig A.F.B. Selma, AL, personnel management; (1951) U.S.A.F. Strategic Air Command 7th BG, Fort Worth, TX, B-36 (10 engine) first pilot check-out training; (1953) U.S.A.F. Air University, field grade, Command and Staff School; (1958) U.S.A.F. Strategic Air Command KC-135 four engine jet first pilot training; (1961) Petroleum Landman's Assn., Minerals Law, Wichita, KS; (1968) State Department, Internal Defense Course, Washington, D.C.; (1974) Treasury, Secret Service Police Academy, Washington, D.C.; (1978) Civil Service, management analysis course, Washington, D.C.; Dept. of Justice-1981, Freedom of Information/Privacy Act course.

Military experience: (1940-60) -1941-flight instructor, Randolph A.F.B., San Antonio, TX; 1942-43- B-17 aircraft commander, combat 8th A.F., England; 1943-44- B-29 service test pilot and operations officer; 1944-45 - B-29 aircraft commander and combat lead pilot, 20th Bomber Command, Saipan, combat tour; 1945-46 - chief of flight test, AMC San Bernardino, CA; 1946-48 service installations officer in charge of post exchange, service clubs, messes, theater, rest camp and post engineers, Muroc, CA; 1948-50 chief of flight test and operations, Alaska Air Depot, Anchorage, AK; 1950-54 tactical operations and training inspector, inspector general, field maintenance squadron commander, chief of Combat Operations, 7th Bomb Wing SAC, Ft. Worth, TX; 1954-55 executive officer, B-26 BG, Laon, France; 1955-57 headquarters section commander, 12th Air Force Commandant, Ramstein, Germany; 1957-60 chief of Combat Operations, wing logistics officer, deputy for Material, 4123rd Strategic Wing, SAC, Clinton Sherman, OK.

Retired from the US Air Force Dec. 31, 1960. Civilian government service: (1968-71), 1966-71 - Field Support Officer, US State Department, Agency for International Development, Vietnam; 1974-1991. Management Analyst, Treasury Department, Bureau of Engraving and Printing. Civilian Work : 1961-66 - Petroleum Landman, Wichita Kansas.

Married to Chi D. Haltom; have two sisters. Retired from Treasury Dept. Jan. 2, 1991. Now managing own properties and wife owns and operates a woman's fashion shop.

Honors and medals: Distinguished Flying Cross with two OLC, Air Medal with three OLC, Commendation Medal with OLC, Purple Heart, European-African Middle Eastern Campaign Medal with two Battle Stars, Asiatic-Pacific Campaign Medal with five Battle Stars, Air Defense Service Medal, American Campaign Medal, National Defense Medal, WWII Victory Medal, Air Force Reserve Medal with Cluster, Air Force Longevity with three OLC, Occupation Medal of Germany, Outstanding Unit Award, Presidential Unit Award with OLC, and Vietnam Civilian Medal.

### JACK GLEATON HAMILTON, M.D.,

Maj., was born March 27, 1918 in Montbrook, FL. Enlisted flying cadet, 1940, from University of Florida with Gordon Gardner, Rudy Miro and Henry Keel at Ft. Scriven, Savannah, GA.

Service in: 4th, 10th, 14th Air Forces and China Air Task Force (July-August-September 1942), 26th F.S., 51st F.G. First assignment in February 1941, 79th Sqd., 20th Pursuit Grp., Hamilton Field, Col. Ira Eaker, group commander.

Planes flown: PT-3, BT-9, AT-6, P-40, P-38, P-39, P-63, P-51, B-25 and C-46.

Memorable experiences: (1) bailout, P-40, India, September 1942; (2) volunteer, co-pilot, DC-3, over "the Hump" instruments, bad weather with inexperienced pilot (luck saved us); (3) mission from Hengyang over Nanchang in August 1942.

Awarded Distinguished Flying Cross and Air Medals. Discharged Jan. 2, 1946.

Post military: M.D., U.S.C., 1951. Urologist. Retired 1983.

Married June 29, 1941, Lanora Ingram, B.A., Stetson University and they have two sons (JGH, Jr. and Thomas Ingram Hamilton) and four grandchildren. Home is in Huntington Beach, CA.

### CLIFFORD HAMMOCK, was born Feb. 21, 1923 in Pitts, GA. Joined Air Force June 10, 1942.

After basic training in Biloxi, MS went to gunnery school in Las Vegas, NV, armament school in Salt Lake City, UT, flight training in Tucson, AZ. Was assigned to 384th BG, 546th Sqdn. as a tail gunner on B-17, *Sad Sack II*.

On 11th mission Sept. 6, 1943, returning from a bombing raid over Stuttgart, Germany, we were hit and were running out of gas. Bailed out near Trie Chateau, France. Had rank of staff sergeant. Received Air Medal with Oak Leaf Cluster. Was gunnery instructor in Gulf Port, MS until discharge Oct. 28, 1945.

Married Frances Bayard in 1947 in Columbus, GA. Moved to Pensacola, FL in 1951. Owned and operated auto repair business until retiring in February of 1985. They have three children.

### MARINO HANNESSON, was born on

Sept. 6, 1915 in Upham, ND. Joined the Army on July 1, 1942 and was assigned to the Army Air Corps. He took basic training and airplane mechanics at Wichita Falls, TX; factory school on B-26s in Baltimore, MD; gunnery school at Buckingham AFB, Ft. Meyer, FL; overseas training at Barksdale Field, LA.

Left for overseas in June 1943 and flew the north route to England over Labrador, Greenland, Iceland and Scotland. Assigned to 452nd Sqdn., 322nd BG, the first mission was Oct. 8, 1943.

Was on 51st mission on May 7, 1944 when shot down by enemy aircraft. Bailed out and landed in the woods. Was helped by the Belgium Resistance and stayed with them until the U.S. 9th Army came.

Returned to England and was sent to the U.S. Spent the remainder of the time at Biloxi, MS as engine mechanics instructor. Discharged Sept. 5, 1945 as technical sergeant.

Married with two sons and one daughter. Retired as a farm laborer in 1977.

### HARRY HANSEN (GUS), Lt. C

U.S.A.F. (Ret.), was born April 4, 1921 Cadillac, MI. Joined the service Dec. 16, 19— Served with 13th A.F., 7th Air Div., 2nd A. SAC. Command pilot with 6,600 hours f— ing time.

Planes flown: P-40, P-47, C-45, C-47, 54, T-28, T-29, B-25 and T-33.

On a routine flight Sept. 19, 1947 fro— Okinawa to Clark Field in the Philippines the northeast tip of Luzon, his C-46 encou— tered a severe line of thunderstorms. T— weather was more severe than briefed to t— pilot at departure and he chose not to g— through the storms. After many turns an— changes of altitude, the plane ran low on fu— and the pilot ordered everyone to bail ou—

It was midnight and the area was ne— Bontoc, Luzon. Hansen landed in a tree hig— in the mountain jungles of Luzon. Aft— three days of struggling through the jungle he was found by a band of Igorots wit— painted bodies carrying spears. They eve— tually took him to a missionary who co— nected him with the rescue team that w— searching for him. He was finally returne— to Clark Field seven days later and te— pounds lighter. Discharged March 30, 196— with rank of lieutenant colonel.

Married Evelyn July 4, 1951 and the— have two children and three grandchildre— Earned a B.A. degree from Omaha, 1965 an— worked for Michigan State Civil Servic— Retired in January 1985 from state civil se— vice administration. Resides in East Gran— Rapids, MI.

### MAE BELLE BRYANT HARDIN, wa—

born Aug. 1, 1921 in Birmingham, AL. Serve— in W.A.A.C./W.A.C. Joined Nov. 28, 1942— Took basic at Ft. Des Moines, Iowa; radi— school at Kansas City, MO; stationed at Perri— Field and Randolph Field, Texas. Served a— an Army control tower operator.

Corp. Bryant was a passenger on a AT— 6 training plane on a cross-country fligh— from Perrin Field, TX to Birmingham, AL t— visit her father. Pilot of the aircraft was L— John W. Hooten. Weather had delayed fly— ing to Alabama, but it was on the return tri— to Texas the two ran into trouble. They ra—

into bad weather and the radio went out. They circled for two hours searching for a hole in the overcast and a place to land. When they ran out of fuel they were forced to bail out at about 7,000 ft. near Hugo, OK. Bryant swung into a high power line and suffered a skinned leg and a cracked rib.

Discharged April 2, 1947 with rank of staff sergeant. Awarded: American Theatre Ribbon, WAAC Service Ribbon, WWII Victory Medal and Good Conduct Medal.

Married Thomas H. Hardin Feb. 14, 1947. They have three children and one grandchild. Resides in St. Petersburg, FL.

She was a draftsman before enlisting and later worked as a church secretary for 20 years. Her sister, Bobbie, also served in the Corp.

Her husband was a B-17 pilot in England during the war, and was in B-29 transition training for a Pacific assignment when the war ended. For many years he was a corporate pilot and only stopped flying a few years ago. He jokingly says he had his feet out the door several times but never had to jump.

Hardin remarked, "Working in control towers at basic flying training fields was never dull, but the most memorable experience was the flight in an AT-6 when I had to walk home."

## HUGH JAMES HARRIES, was born Feb. 10, 1921 in Broadwater, NE. Joined Army Air Force May 14, 1942. Served with 448th B.G., 714th B.S.

Locations: Santa Ana, Ontario, Merced, CA, Ft. Sumner, NM, Alamogordo, NM, Omaha, NE, Newport News, VA and Seething, England.

Memorable experiences: almost being lost before arriving overseas in the Bermuda

Triangle; loss of an engine over the S.A. jungle -exhausted; gasoline at Dakar; loss of an engine over the desert.

Was co-pilot on Lt. Towles crew of the 714th B.S., 448th B.G. March 1944-June 1944, in Seething, England, 2nd Air Div. 8th A.F. on 23rd mission to bomb an air field. At Everevy, France they were hit by an enemy rocket and were immediately on fire in the mid-crew compartment.

The navigator, two waist gunners, nose gunner and Harries were the only survivors in parachuting out of the burning plane. The nose gunner died two days later of burns. A German sidecar motorcycle with two Germans greeted Harries on landing and he was taken prisoner. One waist gunner and the navigator evaded capture for several months, finally being taken in Paris by the Gestapo.

During interrogation, Harries was mistaken for a paratrooper and wound up in a ground officers camp in Poland, then eventually in Luft III in Germany. He made several forced marches in France and Germany. He was eventually liberated by Patton's 3rd Army at Moosberg, Germany.

Achieved rank of first lieutenant and later lieutenant colonel in Reserves. Discharged from active duty in 1945 and from Reserves in 1968.

Attended college for two-and-a-half years. Retired as retirement home administrator in 1981.

Married second wife, Joanne E. Harries, in 1980. He has three children, three step-children, six grandchildren and three step-grandchildren. Resides in Ocean Shores, WA.

Member of: D.A.V., V.F.W., A.F.A., Ex-POW, Luft III, 8th A.F. H.S., 2nd A.D. Liberator Club, MOPH.

## HOWARD M. HARRIS, 1st Lt., was born July 28, 1918 in Port Byron, NY. Joined the Army March 18, 1942. Transferred to U.S.A.A. Corps in July 1942. Graduated class 43-5, April 1, 1943 as a bombardier in San Angelo, TX. Completed crew training at Dalhart, TX. Assigned to B-17's, 100th B.G., 349th B.S. at Norwich, England, Aug. 12, 1943.

On Sept. 3, 1943, on a raid on the Renault Factory over Paris, was hit by FW 190 German fighter planes. Left engine was on fire, tail was shot off, anti-aircraft fire set left wing

tank burning. Bailed out at 28,000 ft. with an Irvin parachute.

Nine of crew parachuted safely. Tail gunner was killed, three were POW's and six managed to evade.

Nearing impact, Harris unbuckled his leg straps for quick getaway. He was shot at while in the air and sustained leg wounds. He landed on the edge of Orly Air Field. Eventually, with the aid of the French Underground, he was able to return to England.

Awards: Purple Heart, Air Medal with OLC, Presidential Group Citation, E.T.O. Medal, American Theatre, Victory Medal, and Korean War Medal.

After returning to States, graduated pilot's Training Class A, Jan. 1, 1945 at La Junta, CO.

Recalled to active duty during Korean War 1951. Assigned to 5th A.F., K5.

Retired from Maislin Transport 1983. Married Jeannette R. Burley on Feb. 22, 1942. They have a son, Gregory H. Harris, and a grandson, Clayton M. Resides in Wolcott, NY

## ROBERT C. HAYNES, M/Sgt. U.S.A.F. (Ret.), was born Sept. 12, 1921 in Ennis, TX. Enlisted in Army Air Corps, Randolph, Sept. 21, 1939. Gunnery school in Del Rio, TX. B-24 training in Idaho and California. Flew from Florida to England with 453th B.G., 733rd B.S.

On Fredickshauen raid, flak knocked out number four engine, forcing them to fall out of formation. The aircraft was hit by FW-190 German fighters. It caught fire and the crew was forced to bail out.

Haynes was captured by Germans. Was also hit in the right arm by flak. Taken to POW Camp VI then Luft IV. Was bayoneted and bitten several times by German dogs on march from boot to POW camp. Walked the approximately 900-mile march

Discharged from U.S.A.F. in 1961. Awarded Air Medal.

## ROGER H. HEBNER, was born March 29, 1910 in Duluth, MN. Enlisted September 29, 1929, 115th Observation Squadron, California National Guard.

Aircraft mechanic, first sergeant. Inducted March 3, 1941. Commissioned

A.A.F.O.C.S., June 24, 1942, stationed Lockbourne, OH, Stuttgart, AR, July, 1944, 434th Sqd., 12th B.G., engineering officer. India. September 1945, commander until deactivation of squadron Jan. 28, 1946 Ft. Lawton, WA.

1946-48, Bomb Wing, Walker AFB, engineering officer, B-29. 1948-50, 33rd F.W., Otis AFB, supply squadron commander F-84, F-86. 1951, Operation Greenhouse, Eniwetok, supply director. 1952-54, special weapons command, logistics liaison (atomic tests) 1954-58, ARDC, ballistic missile division, material director. 1958-61, SHAPE, Paris, logistic plans. 1961 Hq. USAF, Pentagon, logistic plans to retirement, June 30, 1962.

Awards: Legion of Merit, USAF Commendation Medal.

1962-67, aerojet, Sacramento, CA. 1967-74, American River College, Sacramento State University, B.S. and graduate studies.

The 115th Observation Squadron, 40th Div. Aviation was in field training at Camp San Luis Obispo. On the morning of July 12, 1932, they were to make a series of test flights. They took off towards Morro Bay under a ceiling of about 2,500 ft., coastal stratus. They flew into the stratus and started a power dive. The plane whipped violently. Something slapped Hebner in the face and he ducked down in the cockpit. He saw the heels of a pair of boots going upward and out and flipped his safety belt, located his rip cord handle and pushed out himself. He saw the pilot, 1st Lt. Charles W. Haas, the squadron engineering officer, in his chute above him. The plane was lost.

Hebner landed on the upslope of a hill, bounced on the seat of his pants, flipped over and was dragged up the hill. He caught up with Lt. Haas, neither was injured.

Married Evelyn in 1967. They have eight children, 19 grandchildren and 12 great-grandchildren. Resides in Folsom, CA. Hebner and Haas participated in the opening dedication of the Channel Islands Air National Guard Base at Point Mugu, CA, the home of the 146th Tactical Airlift Wing of the California Air National Guard, the successor of the 115th Observation Squadron on Sept. 8, 1990.

## THEODORE W. HERRING (LIGHTNING), S/Sgt. U.S.A.F., was born April 16, 1923 in Newport, AR. Joined Air Force November 2, 1942. Washed out cadet and went to war as engineer/tail gunner, B-24 aircraft, 454th B.G., 15th A.F. based at Cherignola, Italy.

Shot down on second mission. Ditched. Returned to duty and flew 49 missions. On 49th mission to Villach, Austria on Nov. 11, 1944, aircraft exploded while going down before reaching target. Two airmen were

blown free and the other eight were killed in action.

Herring was located and rescued by British Commandos and returned to U.S.A.F. He was hospitalized at the U.S. hospital in Bari, Italy until his return to the States. Herring credits his parachute with saving his life.

Awards: Purple Heart with OLC, Air Medal with Silver OLC,

Presidential Unit Citation, Good Conduct Medal, American Campaign Medal, Euro-Afro-Mid East Campaign with Silver and Bronze Service Stars, WWII Victory Medal.

Married Rudy C. Henry from hometown. They have one daughter and two grandchildren. Resides in Newport, AR.

## CRAWFORD E. HICKS, was born February 10, 1921 in Leitchfield, KY. He was inducted on April 29, 1942. He served with the 351st Bomb Gp. in Polebrook, Northamptonshire, England. He retired from Robins AFB, Warner-Robins, GA. He flew the PT-17, BT-13, AT-10, B-17, B-25, and the AT-6, Cub. He was discharged on January 31, 1967 with the rank of Lt. Col. (0-5) He was married to Rene B. Hicks (deceased) on June 23, 1945. He has three children and six grandchildren. He is currently practicing law and lives in Norcross (Atlanta), GA.

## WILLIAM A. HOFFMAN III, Lt. Col. (ret.), was born Sept. 3, 1918 in Alexandria, VA. Joined Army Aviation Cadets May 10, 1942. After training at Monroe, LA, commissioned second lieutenant navigator and trained for combat at Kearney, NE. Assigned to 92nd BG, flew B-17s from Podington, England.

On Feb. 8, 1944, during sixth mission,

the B-17 was shot down over northern Franc Evaded through France to the Brittany coa with the French Resistance and was picke up by British motor gun boat 503 on the nig of March 23rd and returned to Englan Remained on active duty until Dec. 12, 194

After return to civilian status, work for the Army Map Service as a geodesi retiring March 3, 1979.

Married Ann Bowen of Alexandria. Tw children: Kathleen Cassidy and Ku Hoffman. Four grandchildren: Sabrina, Je nifer and Ryan Cassidy and Keri A Hoffman. Hoffman passed away on Ap 17, 1992. *Submitted by his wife Anna B. Hoffma*

## JACKSON W. HOLLAND (RED), S Sgt., was born March 3, 1924 in Wilson, T Joined the U.S.A.F. on May 28, 1942. Serve with the 9th A.F., 394th B.G., 585th B.S. France.

On Feb. 14, 1945, on his return from h 24th mission from Steinebruck, German the plane encountered intense "88" flak. H draulics and emergency cable were disable on the landing gear. Running out of fue Holland bailed out at 375-450 ft. and lande in a stand-up position. He fractured lumba five-six, losing control of his legs and arm Recovered in the 159th General Hospital i Hereford, England.

Awards: Purple Heart, Air Medal wit three OLC, EAME Campaign Medal wit four Bronze Stars, Distinguished Flying Cros and Good Conduct Medal.

Planes flown: ME 748, B-25, B-26, C-42 B-36 and C-46.

Four of the original crew are still living Discharged Sept. 23, 1945 with rank of sta sergeant.

Married Louise Latta. Had three chi dren, lost one. Has four grandchildren.

Employed by Air Force, General Dy namics, private business. Retired Decembe 1987. Resides in Lubbock, TX. Enjoys trave ing.

## BENJAMIN EGBERT HOPKINS, Maj was born Dec. 21, 1915 in Lincoln County GA. Joined U.S.A.F. Dec. 16, 1941. Serve with 305th B.G., 366th B.S., Chelveston, En gland.

Planes flown: B-17, B-29, C-54 and C-47

Stations: Chelveston, Eng.; Rattlesnake A.F.B., Pyote, TX; Hamilton Field, San Rafael, CA; Hickam Field, Hawaii (Air Transport Command). Separated Oct. 30, 1946. Returned to active duty Nov. 26, 1948. Davis Monthan A.F.B., Tucson, AZ; Chatham Field, Savannah, GA, squadron commander, Hq. and Hq. Sqd., 2nd Air Base Group. Final separation active duty Feb. 26, 1950 with rank of major.

Reserve duty stations: Hunter A.F.B., Savannah, GA; Cheyenne, WY; Dobbins A.F.B., Marietta, GA; plus several 15-day tours in Atlanta, GA and elsewhere.

While flying at 25,000 ft. on May 2, 1945, the left gunner called Hopkins (instructor-pilot) to say number one engine was on fire and the flames were coming back to the horizontal stabilizer. Bailout was delayed due to lack of equalized pressure in the cabin. When corrected, Hopkins gave the order to abandon the plane and was the last to leave. His parachute failed to open when he pulled the ripcord. He managed to pull the shroud lines by hand and capture enough air for three small canopies to open. Hopkins suffered a broken neck, broken left ankle and internal injuries.

Awards: Air Medal with three OLC, Distinguished Flying Cross, etc.

Married Martha March 7, 1944 and has six children and six grandchildren. Attended Jr. College, Berry College. Was a salesman, then postmaster, Washington, GA. Retired July 31, 1975. Resides in Washington, GA. Enjoys family, grandchildren and yard.

## ELMER N HOREY, S/Sgt. U.S.A.A.C.,
was born May 10, 1917 in Fredonia, NY. Joined service Feb. 12, 1942. Served with 392nd B.G., 576th B.S., Wendling, England, E.T.O. Aerial gunner, later navigator on B-24.

Attacked by Me-109 on seventh mis-

sion. Co-pilot was killed and cockpit was set on fire. Nine bailed out between Frankfurt and Hanover, Germany. Horey broke his ankle in the parachute landing. He spent 13 months in Stalag 17-B Krems, Austria.

Awarded: Purple Heart, POW Medal, Air Medal, American Campaign, European-African-Mid East Campaign, WWII Victory Medal and New York State Conspicuous Service Medal.

Married Bertha and has one son and three grandchildren.

Retired school teacher and school administrator. Earned B.S. and M.S. in education. R.S.V.P. commander, Ex-POW, southwestern New York chapter.

Discharged Oct. 30, 1945 with rank of staff sergeant.

## DORAN HUGHES (GREMLIN), Sgt.
was born May 4, 1925 in Raisinville, MI. Joined the A.A.F. August 13, 1943. Completed aerial gunnery training at Las Vegas aerial gunnery school and was assigned as tail gunner to a B-17 crew training at Pyote, TX, 8th A.F.

Was injured in a parachute jump on June 10, 1944 over Abilene, TX. B-17 was on fire, right wing burning and fuselage was full of smoke. Order was given to "Abandon ship!" Hughes reluctantly went head first out of the tail gunners hatch. He was turned end over end and says he will never forget the jolt when "that little chest pack brought me to a complete stop". The whole crew got out safely but Hughes landed on a moving automobile and knocked three vertebras out of place in his back when the chute became tangled and dragged him beneath the car.

Later reassigned to Grand Island, NE, for familiarization training as a relief gunner on the B-29 Superfortress. Was sent to South Pacific and served with the 6th B.G., 20th A.F. on Tinian Island. Discharged Jan. 8, 1946.

Awarded the Bronze Star, Air Medal, Asiatic-Pacific Theatre Medal, American Campaign Medal, Japanese Occupation and WWII Victory Medals.

Married Theresa (Terry) Arbic on July 18, 1946 in Sault Ste Marie, MI and moved to Kenosha, WI. Has two sons and four grandchildren. Retired as Kenosha County,

Wisconsin director of emergency services on Jan. 16, 1987.

## JEROME JACOBS, was born Oct. 26, 1923 in U.S.A. Joined U.S.A.A.F. Dec. 7, 1942. Served in England with 364th F.S., 357th F.G.

On approximately Sept. 15, 1944, Jacobs was the pilot who spotted the German battleship Tirpitz in a fjord in Norway. He reported the sighting which was verified by a British recon aircraft on the same afternoon which was sent out to verify his sighting. The Tirpitz was subsequently sunk by British Lancasters on third attempt. On 16th mission, Sept. 18, 1944, operation "Market Garden", Jacobs shot down one Me-109 and one FW-190. On 17th mission, Sept. 19, 1944, same operation, shot down a Me-109, five minutes later was shot down and bailed out. Captured immediately and was in prison camp until ten days before VE Day. Liberated by Patton's 3rd Army.

Most memorable experience: three air-to-air victories (two Me-109s and one FW-190) on same mission. Discharged in January 1946 with rank of first lieutenant.

Married with two children and three grandchildren. Oral surgeon.

## ALBERT W. JAMES, was born April 27, 1916 in National City, CA. Joined Army Air Corps/U.S.A.F. Sept. 21, 1939.

Military locations and stations: Randolph and Kelly Fields, Hamilton, Salt Lake, Manchester, NH, Mitchel Field, Miami, Philippines, Dutch East Indies, Felts Field, Dover DE, Wright-Patterson, NY, Koblenz, W. Germany, Patrick A.F.B., Langley Field.

Served with 7th B.G., 45th B.G., 19th Anti-sub Sqd. 459th B.G., Aeronautical Systems Center, 5th B.G.

Bailout occurred April 4, 1941. James was on a mission to deliver his friend, Lt. Jack Alston's A-17, to Denver, CO from the Salt Lake City area. The trip to Denver was uneventful, but on his return to Salt Lake he ran into unforcasted bad weather and trouble with the instrument panel. Cloud cover and darkness made it impossible to locate any of the auxiliary fields he was searching for. Unable to stabilize the plane, James elected to bail out. He had no glimpse of the ground or warning of its approaching due to the cloudiness and darkness before striking it. Fortunately he landed in soft ground and suffered only a black eye. The young man had no idea where he was except for knowing he was somewhere in the Rockies near the top of a ridge. He spent that night and the next day making his way back to civilization and was finally picked up on a highway by a local man, William Keys from Taggart's Canyon Camp, a few miles east of Ogden, Utah.

Aside from commanding the 5th B.G., 13th Air Force FEAF, James stated that his most memorable experience was as liaison to four nations (West Germany, Italy, Netherlands and Belgium), for the F-104 Consortium program.

Educated at the Air Force Institute of Technology, M.B.A. at University of Texas. Retired from service Oct. 31, 1965.

Married Mildred H. James Dec. 31, 1965 (2nd go). Has three plus six children and nine grandchildren.

Worked for Boeing-Apollo and Minuteman Management. Retired May 31, 1976. Resides in Melbourne Beach, FL.

## ALEXANDER JEFFERSON, Lt. Col., U.S.A.F. (Ret.), was born Nov. 15, 1921 in Detroit, MI. Joined Air Force Sept. 18, 1942. Served with 15th A.F. in Italy, 332nd F.G.

On Aug. 12, 1944, on 18th mission, he was shot down while strafing radar stations on the coast of southern France at Toulon, prior to the invasion. He was hit by 20mm shell that came up through the floor just in front of the stick, out through the top of canopy. Pulled up and bailed out at about 800 ft. Chute opened while going down through the trees. Landed in the middle of 20mm crew.

Awarded Air Medal, ETO Medal, American Theatre Medal, POW Medal, Victory Medal, Air Force Reserve Medal.

Retired from Air Force Reserves 1969. He has a masters degree in education and retired as assistant principal, Board of Education, City of Detroit, 1979. Has two daughters.

Life member of Silver Falcons; perpetual member of Military Order of the World Wars; Air Force Assoc.; Tuskegee Airmen, Inc.

## OTIS E. JOHNSON, T/Sgt., was born March 6, 1918 in York, AL. Joined U.S.A.C./U.S.A.F. in 1939, 1940 and 1950.

Served in U.S., Egypt, Jordan, Libya, Sicily, British Isles, Holland and Germany. Served with Mississippi National Guard, 1st Div., 16th Inf., 316th G.P., 36th B.S.

Shot down while towing gliders over Holland and with help, hid from the Germans. POW.

Discharged January 1966 with rank of technical sergeant. Education: high school, B.S., Economics, Psychology, plus graduate courses.

Married and has two children and one grandchild. Worked for U.S. Civil Service as education officer, State Civil Service as case worker. Retired in 1983. Resides in Temple Hills, MD

Remark: "My unit, 36th Sqd., was involved in nine campaigns, N. Africa and the E.T.O. They also received three Presidential Citations. We lost quite a few good men."

## LAWRENCE JUSTIN, (LARRY), Capt. U.S.A.F., was born Oct. 29, 1916 in Burlington, VT. Enlisted April 1941. Commissioned pilot class 42C Maxwell Field, Montgomery, AL. Attended Fighter Control School, Orlando, FL, then ordered to Uxbridge, England as liaison controller, R.A.F. Fighter Command, 11th Group. September 1943, assigned to Hq. 9th Fighter Command, at Middle Wallop; May 1944 assigned to 367th F.G. at Stoney Cross. Flew P-38s, P-47s, P-51s.

On June 22, 1944 was hit by flak over Cherbourg and parachuted into English Channel. Picked up by British Air-Sea rescue and returned to base unharmed.

Most memorable experiences: flying cover over Omaha Beach June 6, 1944 (D Day) and observing thousands of boats and troops involved in landing; flying ground support to Gen. Patton's 3rd Army in advance across Europe.

Awards: Distinguished Flying Cross, Bronze Star with V, Air Medal, Presidential Citation, ETO Medal with seven Battle Stars. British awards: 1939-1945 Star, Aircrew Europe.

Flew 87 combat sorties. Discharged December 1945 with rank of captain. Married to Norma Aug. 11, 1940 and has two children

and four grandchildren. He retired from the construction business and resides in Surfside FL.

## VICTOR J. KARASEK, was born July 18 1915 in Montreal, Canada. Education: high school, Hill College, Quartermasters school Camp Lee. Joined Quartermasters Sept. 17 1941. Served at Camp Lee, Kelly Field, Lake Charles, LA and Greenville, TX with Det. 90: QM., 1144 QM., 320 Bombardment. Achieved rank of master sergeant.

On July 23, 1944, while flying from Elaounina Air Base in Tunisia to Italy the signal to vacate the plane was given when the B-17 caught fire. All were wearing Mae Wests, but chutes were being passed to the rear. The door in the side wouldn't open so the crew started to push through. Karasek was one of the last, being first sergeant.

He pushed out of the door and all the crewman could remember was a shot going off. His chute had opened and when Karasek looked back he could see the others opening. He finally landed in the water and discarded his chute. He heard cries for help and swam to help but could not locate the man.

Karasek floated in the Mediterranean for almost 24 hours before swimming to shore near Cape Bon.

While going through the mountains he ran across an Arab who spoke to him in German. Karasek did not know German so he spoke to him in French. He asked if there were any military installations in the vicinity and the man said no. The airman asked for bread and water and he brought him to a cave and told him he would return with water and bread. The man finally came back with the bread and water and said he would return when it was dark . Karasek watched him leave and followed him to a French outpost on top of the mountain. The French took him in and called the authorities at the base. The next day he was taken to the police station where he met another one of his men. They were taken to the French Embassy where an ambulance took them to the hospital.

Twenty-two men from the 1144 QM Co. leaped to their deaths and the crew chief died also on that fateful morning, July 23, 1944. The irony of it was that the plane landed back at the base with seven men aboard.

Karasek was discharged Dec. 15, 1945 with the rank of master sergeant. He married and has two children and three grandchildren. He retired Dec. 31, 1979 from Draper Corporation where he had been a set up man. He now resides in Blackstone, MA.

**PAUL P. KASZA,** was born Jan. 14, 1921 in Cleveland, OH. Entered service Nov. 3, 1942. Graduated radio operator mechanic school Chicago, IL May 1943. Graduated aerial gunnery school L.V.A.G.S., Las Vegas, NV July 1943. Assigned to E. Fitzpatrick crew Aug. 1943. Trained on B-24 at Davis Monthan, Tucson, AZ and Blyth, CA.

Arrived in England Nov. 1943 assigned to 8th A.F. 801st BG, 406th BS at Alconbury, Watton and Harrington, made rank of technical sergeant.

Flew first carpetbagger mission Feb. 6, 1944. Shot down 1:30 a.m. by German Me110 night fighter on 17th carpetbagger mission May 30, 1944. Evaded (Belgium) May 30, 1944 to Oct. 2, 1944. Returned to the 492nd BG and departed for U.S. Oct. 22, 1944.

Awards: Distinguished Unit Citation, WWII Victory Medal, Air Medal with two OLC, E.T.O. Ribbon with two Bronze Stars, Purple Heart, Good Conduct Medal.

Married, two children. Retired carpenter, 1984. Lives in Seven Hills, OH.

**NORMAN P. KEMPTON,** was born May 13, 1918 in Morristown, SD. Obtained a private pilot's license in 1939 through the civilian pilot training program at South Dakota School of Mines, Rapid City, SD. Joined the U.S. Army Air Corps Aug. 20, 1940. Graduated from the U.S.A.A.C. Aircraft Mechanics School at Chanute Field, IL in 1940.

Military pilot training was completed at the Southeast training command class 43C at Maxwell Field, AL, Orangeburg, SC and Columbus, MS. B-17 transition training was at Sebring, FL in 1943.

After crew training, he was assigned to the 8th A.F., 388th BG, stationed near Thedford, England.

Shot down on 10th mission Jan. 4, 1944, near Bordeaux, France. Parachuted and

evaded capture for ten days until contact was established with the French Resistance. Walked over the Pyrenees and reached Isaba, Spain on Feb. 23, 1944. Returned to England April 13, 1944 by way of Gibraltar.

Next tour of duty was piloting C-54s with the North Atlantic Div. of the Air Transport Command.

Discharged November 1946. Obtained a B.S. degree from Colorado State University in 1949.

Upon receiving degree, was employed as a research and development engineer at Wright-Patterson A.F.B., OH and later as an aerospace engineer at Patrick A.F.B., FL. Retired after 36 years with the Air Force in December 1979.

Married Pearl L. Sturgeon and they have one son, Gary L. and one grandson, Wesley Norman.

**EARL LEHMAN KIELGASS,** born in Miami, Arizona. Pilot training was with class 43-K (Santa Ana, CA; DOS Palos, CA; Lemoore, CA; Williams Field, AZ;). P-47 Thunderbolt RTU at Dover, DE. Sent to United Kingdom in May 1944 and after a short period at Atchem, joined 368th F.G. at A-3 near Isigny, France on July 1, 1944. Assigned to 396th F.S. and stayed with it for remainder of war against the Reich (other European WWII bases were: Chartre, Laon, Chievre, Rheims, Metz, Frankfurt, and Nuremburg).

Separated from service in 1946 after serving as adjutant of pilot students at Luke and Williams bases. Joined Arizona National Guard in early 1947 and flew P-51 Mustangs at Luke until recalled to active regular Air Force in the fall of 1947. Was assigned to 1st Fighter Wing, March Field,

until he wrangled a Berlin Air Lift assignment in early 1949.

Went through C-54 pilot transition at Great Falls, MT, then joined 53rd Troop Carrier Sqd. at Rhein Main and hauled supplies into Berlin for the next several months. Spent several years in USAF Training Command (Enid, OK; Big Springs, TX) with duty as pilot training group and wing adjutant as well as thousands of hours as an instructor pilot in AT-6, C-47, B-25, AT-7, T-28, and T-33 aircraft. Was commander of pilot training sections I and III for about two years at Webb A.F.B. Was selected by USAF Hq. for a SHAPE Hq. assignment and spent 39 months at Villacoublay (outside Paris) with air training advisory group which had the NATO responsibility for all pilot training within the Alliance.

In 1959, he joined A.A.C.S. after attending the air traffic control staff officer course at Keesler A.F.B. In A.A.C.S. he was squadron commander (Whiteman A.F.B.) communications region inspector general (Chanute A.F.B.), and also spent two years in southeast Asia as director of air traffic control services for the SEA areas. In 1965 Kielgass was detailed by USAF to the F.A.A. in Washington, D.C. and served three years with research and development division as program manager for A.T.C. Systems. In 1968 was assigned to United Kingdom command region (3rd A.F. South Ruislip) as director of U.S.A.F. A.T.C. services. He stayed in England for four years then came home to Williams Air Force Base in 1972 to retire. Married Loraine Jenkins from Central City, KY on Dec. 14, 1942. They have two children, Dennis and Susan. Kielgass attended Arizona State University, Tempe, AZ and earned a BA and MA in language and literature. Resides in Tempe, AZ. Joined ranks of those who owe their lives to the wonderful parachute at 1015 AM 12 March 1958 near Wiesbaden, Germany after ejecting from a burning T-33.

**ROBERT J. KIERNAN,** was born April 12, 1922 in Akron, OH. Joined the service Oct. 12, 1942. Served in the Army Air Corp as navigator - 1034, with the 305th B.G., 366th B.S., 8th A.F. Kiernan bailed out over the Baltic Sea. Two survived. Was a POW for 11 months at Stalag Luft III Sagan, Germany, Stalag XIII D Nurenburg, Stalag VII A Moosberg. Liberated April 29, 1945.

Awarded Air Medal with two OLC, Purple Heart with OLC and POW Medal.

Kiernan married and has one daughter. He graduated BEE George Washington University, Washington, D.C. Operates a MARS/Navy-Marine radio station. Retired from Dana Corporation.

Discharged from service with rank of first lieutenant. Resides in La Habra Heights, CA.

**EMIL M. KLYM,** Lt. Col., U.S. Army (Ret.), was born Jan. 26, 1918 in Chicago, IL. Inducted Nov. 2, 1942. Served in England with 8th A.F., 447th B.G. (H), 711th B.S., flight engineer and gunner B-17 aircraft. Transferred to 9th A.F., Aug. 6, 1944. Served in England and France with 323rd B.G. (M), 455th B.S. as flight engineer and gunner on B-26 aircraft. Also served in Germany, Greenland, and Japan.

Most memorable experience: During the Battle of the Bulge, his group's mission was to fly over the German troops at an altitude of 100 ft., strafe and drop 100 lb. anti-personnel bombs if the German army advanced beyond a designated point in their drive to take Antwerp, Belgium. They were put on standby and spent most of the day on the flight line awaiting the order to take off. The flight crews were very quiet and kept their thoughts to themselves until the news was received that the mission was canceled. At that time they broke into smiles and there was laughter and chattering. Klym said he had often wondered what the outcome would have been and how many of them would not have returned if the mission had not been canceled.

Klym was forced to bail out Feb. 10, 1945 on a bombing mission to Cologne, Germany because his aircraft was severely damaged by German flak. He landed in some trees, breaking his back, and was aided by Belgiums.

Discharged Oct. 19, 1945 with rank of technical sergeant. Recalled during Korean Conflict. Retired Feb. 24, 1969.

Awarded: Air Medal with five OLC, Purple Heart, ARCOM with OLC, E.A.M.E. with five Battle Stars, G.C.M., A.F.R.M., N.D.S.M., and WW II Victory Medal.

Klym is divorced and has one son, Gary M. Enjoys travel and golf. Resides in West Frankfort, IL.

**A.D. KNEALE, JR. (DELL),** was born Dec. 17, 1916 in Tulsa, OK. Enlisted in Oklahoma Air National Guard, 125th Observation Sqdn., Tulsa, in April 1941. Unit was called to active duty in September 1941; applied for and was accepted in Aviation Cadet training in December 1941. Received navigation training at Hondo, TX navigation school.

In January 1943 was assigned to 551st Bomb Sqdn., 385th BG, Great Falls, MT. Flew with this unit in the E.T.O. as navigator on the *Old Shillelagh* and *Old Shillelagh II*.

On fifth mission, ditched in the North Sea and was picked up by Air Sea Rescue. On eighth mission, Aug. 25, 1943, shot down over France and broke his leg after bailing out. Evaded capture and walked over Pyrenees.

Arrived back in U.K. after 67 days. Returned to States and taught navigation at Hondo and later radar at Victorville, CA. Separated from service as first lieutenant October 1945.

Awarded Purple Heart and Air Medal with Oak Leaf Cluster.

Now retired after various accounting positions including 23 years with Sun Oil Co., and ending with three years in his own C.P.A. practice.

Married to Geneva and has three children and three grandchildren. Resides in Bella Vista, AR.

**STANLEY J. KNOTOWICZ,** was born Oct. 23, 1922 in Rochester, NY. Joined Army Air Force in December of 1942. Served with the 15th Air Force in Italy.

Bailed out on June 26, 1944 over Austria on 28th mission. Taken POW. Was POW until May 3, 1945. Was on the 800 mile death march, February 1945 to May 1945.

Awarded: Air Medal with OLC, POW Medal, and Good Conduct Medal.

Married to June, his wife of over 48 years. They have three children, six grandchildren and two great-grandchildren.

Retired from Eastman Kodak Company after 28 years of working there.

Discharged with rank of staff sergeant. Resides in Naples, NY.

**PETER KOUZES,** was born in Batavia, IL. He has four sisters and five brothers. All of his brothers were in the service. (Tom Lt. Col., Nick - master sergeant (30 yrs.), George -technical sergeant, Gust - U.S.N., and James (Intelligence) A.F.).

Kouzes joined the service Jan. 4, 1943. Served with two ATS Mobile, 20th A.F. Was a technical sergeant/radio operator flying - #756 MOS. Flew C-46, C-47 and C-54. Forced

to bail out of a C-46 Commando on Dec. 1 1944 carrying a load of aviation gasolin from Luliang, China to Lahoko. They g lost and were attacked by three Japanes fighter planes and one biplane. The cre was ordered to bail out after the right e gine was hit. Kouzes was soaked with flan mable liquid but managed to jump out th cargo door. Seconds later the aircraft turne 180 degrees and exploded, spewing burr ing metal on the descending crew. An e emy plane headed straight for Kouzes ar the airman prayed and then fell limp pr tending to be dead. His ploy worked. Thre of the four crew members survived.

Awarded: Air Medal with two OL( and Distinguished Flying Cross. Discharge October 1945 with rank of technical se geant.

Semi-retired real estate broker and no part time actor/singer for dinner theatre Resides in Elkhorn, WI.

**ALAN KRONSCHNABEL,** Capt., wa born March 30, 1923 in Dickinson, NL Joined the service in June of 1943. Gradu ated pilot school Stockton, A.F.B., CA Apri 1944 as second lieutenant. Transitions to B 17 Roswell, NM A.F.B. May 1944-Augus 1944.

Assigned combat crew Biggs Field, E Paso, TX August 1944 to October 1944. As signed to 447th B.G. (H), 710th B.S Rattlesden, England E.T.O. in Novembe 1944. Hit by flak and downed on Jan. 2 1945 over and near Verdun, France during Battle of the Bulge. Tail gunner was lost t enemy action due to premature bailout Armorer was badly injured in bailout bu survived. All other crew bailed out safely

All were picked up by F.F.I. (Maquis

and delivered to nearby 9th A.F. Fighter Group. Returned to base in England in February of 1945.

Flew 22 more missions prior to the end of war in the European Theatre. Reassigned to B-29 transition Drew Field, Tampa, FL June 1945.

Released from active service in December 1945 with rank of captain U.S. Air Corps.

Currently semi-retired real estate broker. Married with three children and two grandchildren.

Awarded: Air Medal with four OLC, Presidential Unit Citation, World War II Victory Medal, E.T.O. and Ardennes Campaign Ribbons.

Resides in Lake Chelan, WA.

## ALVIN RAYMOND KUBLY (RAY),

was born May 22, 1924 on the Kubly farm in Watertown, WI. As a youth, he worked on the farm, was active in 4H work and served as basketball manager for his high school team. After graduating in 1942, he enlisted in the Air Force on Oct. 7, 1942. Graduated as a bombardier-navigator and was assigned to a B-17 Flying Fortress with the 34th BG in the 8th Air Force.

On his ninth mission, an oil refinery target near Meresburg, Germany, the plane was hit by anti-aircraft flak and the crew had to parachute out over Holland. He was shot through the leg while still in the air and when he landed the Germans took him prisoner.

Germans took him to the St. Antonius Hospital in Utrecht, Holland and after 19 days there, he and five other P.O.W.s escaped. The Dutch Underground took care of them. After five months and 23 days, he returned to freedom through the front lines.

After WWII, he returned to the family farm until May 1948. He married Ruth Wegwart Sept. 4, 1948. Graduated from the University of Wisconsin-Madison in 1952 with a B.S. in Soils and Agronomy and worked for Cargill Seeds in Rochester, MN. A son, Roger Ray was born July 19, 1950. Joan Kristine was born June 15, 1952 on Father's Day. Carol Ann was born March 30, 1955 in Tracy, MN. Mary Lee was born Jan. 6, 1960 in Watertown, WI.

Was in Air Force Reserve and in 1968

was promoted to lieutenant colonel. Went to work for the Dairyland Seed Co. in 1961 as only salesman. By 1979 he was vice president of sales with over 100 employees now in the company.

Served on Watertown School Board, president three terms. Also served five years on the CESA Board of Control, three as chairman. In 1987 Lt. Col. Kubly was selected as executive sec./treasurer of the Reserve Officers Assn. and as editor of *Wisconsin Reservist*.

Ray and Ruth Kubly have seven grandchildren.

## VINCENT LAYBE, was born July 19, 1919
in Clayton, NM. Was in Arizona National Guard 1939, transferred to Air Force in 1941, 99th BG, 348th Bomb Sqdn. Stationed at Scottfield, McDill, Gowen, Sioux City, Tyndell Field, Mitchell, SD, and overseas. Achieved rank of technical sergeant.

His B-17 was severely damaged by flak Sept. 2, 1943 on 19th mission, target Bologna, Italy. Crew bailed out and was captured by the Italians and sent to Camp 54, Sabina, Province of Rome. Escaped and headed for hills. Lived behind enemy lines for 10 months in village near Rome and in Rome, aided by underground. Lived in different places until our forces took Rome in June 1943.

Returned to base in Foggia, Italy, then sent to a hospital in Bari, Italy and finally stateside to a hospital in Washington State. Discharged in 1945. Entered A.S.U., Tempe, AZ. Met and married Patty Kielgass and had two sons, Michael and Patrick. Lost Patty to cancer in 1988.

Retired after 33 years of teaching science. Enjoy tennis, science projects and astronomy.

Awards: Air Medal, Purple Heart, P.O.W. Medal. Resides in Phoenix, AZ.

## ELMER T. LIAN, was born April 17, 1918
in Fairdale, ND and graduated from Fairdale High School in 1936. Attended the University of North Dakota, graduated in 1940 and received a commission in the U.S. Army as a second lieutenant from the Reserve Officers Training Corps.

Entered the Army and was assigned to

the 3rd Inf. at Ft. Snelling, MN. Joined the Army Air Force in 1942. He retired as a lieutenant colonel on Nov. 1, 1964 at Hamilton AFB, CA.

Married Edwina Amundrud, Fairdale, ND at Ft. Dix, NJ on July 16, 1942. Linda was born July 2, 1944 at Grand Forks, ND and Steven was born Sept. 22, 1947 at Greenville, SC.

## GARY L. LOCKS, was born June 26, 1943
in Passaic, NJ. He graduated from Grove City College, PA January 1967. Commissioned a 2nd lieutenant through R.O.T.C., assigned to 17th BW and went TDY to Da Nang in 1968, where he went on an 0-2 mission and was shot down over enemy territory. He escaped after 17 days and returned to kill his nine captors with an Army patrol who found him on the night of his escape. He returned to Wright-Patterson Air Force Base, OH.

In 1972, he married Deborah Long of Beavercreek, OH. He went to Wright State University and finished his Masters program in guidance counseling.

He is in the active Reserves as a lieutenant colonel. He is assigned to Defense Logistics Agency (DLA). His full time job is in civil service on the B1 Bomber program in logistics management at WPAFB, OH.

Locks is a member of the American Defense Preparedness Assn., Society of Logistics Engineers, National Contract Management Assn., American Legion, The Company of Military Historian, V.P. of the Kittyhawk Chapter, Reserve Office Assn., Young Republicans, NRA, and American Ex-Prisoners of War.

## ALEXANDER MacARTHUR, was born
Oct. 1, 1922 in Illinois. Education: Lake Forest Academy, Hunn School, College of William and Mary.

Joined U.S.A.A.F. Oct. 5, 1942. Served with the 15th A.F., 98th B.G. (H) in Lesse, Italy.

Downed and taken POW. Imprisoned at Stalag Luft III, Stamlager XIII-D and Stalag VII-A.

Discharged in November 1972 with rank of colonel.

Married to Barbara H. MacArthur. They

have six children and five grandchildren and reside in Barrington Hills, IL.

Chairman of Illinois Racing Board and Illinois State Fair Board. President of Illinois Sheriffs Merit Commission Board.

## EDWARD F. MANSFIELD (NED), was born Oct. 23, 1918 in Los Angeles. Joined the service March 15, 1941. Served with the western flying training command.

Planes flown: B-13, AT-6, AT-17, B-25, P-38.

Qualified to be a Caterpillar while flying a BT-15 above the Cascade Mountains in Oregon. Ran into a non-forecasted snow and ice storm. Plane iced and antenna broke off. Radio contact was lost. Jumped at 8,000 ft. Plane did not burn and personal property was taken out of it after spotters saw the plane come out of overcast.

Awarded Air Medal, American Theatre, etc.

Discharged April 15, 1946 with rank of lieutenant colonel.

Mansfield and his wife of over 48 years, JoAnn, have three children and seven grandchildren.

He is now retired and does volunteer work. He enjoys tennis, golf and woodworking.

## JOHN M. MARR, was born Aug. 13, 1921 in Columbus, IN. Graduated from Columbus High School 1938 and Kentucky Military Institute 1939. Attended Purdue University three years studying mechanical engineering. Enlisted Army Air Corps in 1942 and spent four years in military service.

Trained in Texas, Oklahoma, Kansas and at Dale Mabry Field, Tallahassee, FL. P-40 fighter plane training at Thomasville,

GA. P-51 pilot in 8th A.F. based at East Wretham, England.

Shot down over France Jan. 29, 1945, bailed out, hit the airplane tail, spent 20 months salvaging leg. (Revisited France 1985 and located town in whose main street he came down, people who remembered him, and dug up pieces of his airplane that crashed in woods.

Thirty years in motel business in Tallahassee. Retired 1976 to spend more effort promoting constitutional government as a conservative, political activist.

Marr is a Christian worshipping at First Christian Church. He is a member of D.A.V., American Legion, The Reserve Officers Assn., The John Birch Soc. and Rotary.

Family is wife, Norma, Jeff with two sons and working with Postal Service in Brentwood, TN, and Greg with one son and working as a manager with the Florida Department of Revenue, and Donah with two daughters and one son, a homemaker in Tacoma, WA.

Commercial pilot, holding a flight instructor's certificate in airplanes and instruments, but doesn't have time to practice it. Has a real estate salesman's certificate. Teaches an occasional C.P.R. course and enjoys that along with yard work, reading, traveling, visiting grandchildren and church work.

## WILLIAM E. MARTIN (BILL), was born Jan. 25, 1920 in Shelly, ID. Joined the USAF Nov. 18, 1941, assigned to 8th AF, 384th BG, 547th Bomb Sqdn. Stationed at Grafton Underwood, England. He flew the B-17 (Patches). Discharged Sept 5, 1945 with the rank of T/Sgt.

Married Tyleen on Oct. 24, 1944. They have two children and four grandchildren. Worked for Mountain Bell Telco for 43 years. Retired January 1981. His hobbies are stained glass, travel, and woodworking.

## FRANK F. MARVIN, Lt. Col. U.S.A.F. (Ret.), was born May 26, 1923 in Honolulu, HI. Was a cadet in the U.S.N.A. July 1, 1942. Was a second lieutenant in the U.S. Army Air Corps. Served with the 414th F.G., 18th F.G., Clark A.F.B., Philippine Islands.

Planes flown: P-40, P-47, P-51, training PT-19, T-6, T-28, C-45, B-25, T-33.

Shot down March 15, 1948 in a P-47, on of two in what was supposed to be a fo ship combat mission. Everything was goir along normally until Marvin's engi stopped abruptly and the propell windmilled at 18,000 ft. He lowered the no and held an air speed of about 150 mph. F called Lt. McCuff in the other plane to t him of the emergency as he continued pe forming the necessary emergency proc dures. He smelled gas fumes in the cockp and immediately opened the canopy, turne off the emergency booster pump and cut h mixture. He was at about 13,000 ft. and w. headed directly for Clark Field, five mil east of Tarlac. When he realized he had con plete engine failure, he started to prepare fo bailout. An explosion engulfed him in flame but he remained conscious and released h safety belt and bailed out the right side. H did not release the canopy, cut the switch or unfasten any equipment.

His chute opened at about 10,000 ft. an the descent took approximately 10 minute He landed in an open plowed field and su fered a bleeding head wound. Helped b natives and taken to the Philippine Arm Hospital at Tarlac.

Awarded: two Commendation Meda Earned M.S. Aero Engineering from Air For Inst. of Technology.

Retired August 1969 with rank of lieu tenant colonel.

Married Shirleigh McD. Marvin in O tober of 1953. They have three sons.

Retired for the second time after 18 yea on faculty of Virginia Polytechnic Institut and State University, College of Enginee ing.

## FRANCIS C. MARX, T/Sgt., was bor March 1, 1918 in Oppenheim, NY. Entere service Nov. 2, 1942, Scott Field, IL, Laredc TX, Clovis and Alamogordo, NM. Basec Seething, England, 8th A.F., 713th Sqdn 448th BG.

Returning from a mission t Ludwigshafen bailed out north of Paris. / Parisian family hid him until he was directe towards neutral Spain. As he walked ove the Pyrenees, guides deserted him in a freez ing rain and he climbed on while assisting a injured British pilot. Reaching Spain he slep on the floor of jails until rescued by the American Military Attache and became an internee.

Upon return to U.S., he was named N.C.O. in charge of Intelligence, 5th Service Command, St. Louis. Discharged Nov. 13, 1945.

Married Ida Mae Wiener Sept. 29, 1943. Children: Francis C. Jr., Sioux Falls, SD; Dean L., Brussels, Belgium; Michelle Ann Gardiner, Cresskill, NJ. Grandchildren: Brian, Michelle, Kelly Ann, and Ryan Francis.

Retired accountant/auditor and wife Ida is retired school educator, both Rider College. Residence Tampa, FL.

## ARTHUR W. MATTSON (MATT), T/Sgt.,

was born April 4, 1920 in Sidney, MT. Enlisted Nov. 13, 1940. Served with 743rd B.S., 455th B.G., 15th A.F.

On the Munich Mission June 9, 1944, his B-24, the famous "Leakin Deacon" was crippled by flak. Unable to return to Cerignola, pilot headed for Corsica. They cleared peaks by only ten feet while losing altitude through Austrian Alps. Also flew into Deadends.

Descending to 500 ft., the crew bailed out over Guastalla near River Po. Radio operator was killed as his chute failed to open. Five were captured. Nose gunner, top gunner, co-pilot and Mattson evaded capture.

Moving south from farm to farm, they joined the mountain Partisans and passed through German lines north of Florence on Oct. 24, 1944.

Awarded: Air Medal, European Theatre Medal and Pre Pearl Harbor Ribbon.

Discharged Oct. 17, 1945.

Six crew members attended the 1986 Dayton Caterpillar reunion.

Manager of import-export department in Houston, TX for 30 years. Retired in 1982.

Married to Lucille and they have a daughter, a son, and three grandchildren.

## ROY O. McCALDIN,

was born in Shreveport, LA. Joined the service Dec. 12, 1942. Served with the 305th B.G., 8th A.F. Flew B-17.

How qualified to be a Caterpillar: straggler going in to Berlin (March 18, 1945), gun

fight with MeZGZ. He won. Popped chute open in airplane but worked anyway.

Discharged in May 1946 with rank of second lieutenant.

Awarded Air Medal and European Theatre Medal.

Married to Dede and they have three grown children.

Enjoys building and flying airplanes.

## SAMUEL MELANCON, S/Sgt.,

was born Feb. 2, 1916 in Donaldsonville, LA. Joined the military Oct. 10, 1941.

On 36th mission, Feb. 22, 1944 was shot down over the Netherlands above enemy territory. He was knocked unconscious and Norman Bell pulled him out of the tail gun section. He put on Melancon's chute pack and pushed him out of the waist window. He landed in Breda, Holland.

Major Thornton was the pilot and they were leading a flight. After days of walking, received help and was taken to a convent near Antwerp, where a doctor treated him for frozen feet. Later helped by the "White Brigade" who took him to Mr. Albert Van Campenhout's home where he remained for six months until liberation.

After returning to New Orleans, kept in touch with benefactors by corresponding and returned to Belgium several times. Passed away May 21, 1990. *Submitted by his widow, Blanche Melancon.*

## RANDELL S. MEYER,

was born Dec. 6, 1956 in Covina, CA. Entered U.S.A.F. Academy June 30, 1975. Graduated May 30, 1979. Commissioned as second lieutenant.

Unit/Group(s) served military station/location(s): Williams A.F.B., AZ (December 1979-March 1982) 82nd Field Maintenance

Squadron, maintenance officer, 82nd Flying Training Wing, wing job control officer, wing assistant maintenance control supervisor, undergraduate pilot training. Hill A.F.B., VT (January 1983-January 1986) 421st Tactical Fighter Squadron. MacDill A.F.B., FL (February 1986-October 1988) 72nd Tactical Fighter Training Squadron - instructor pilot, 56th Tactical Training Wing - weapons officer. U.S. Embassy Copenhagen/Skrydstrup R.D.A.F. Base, Denmark (October 1988-November 1989) 727th ESK Squadron - F-16 instructor pilot, Embassy officer, flight examiner, flight commander. Ft. Stewart, GA (December 1989-present) Detachment 2, 507th Tactical Air Control Wing - brigade air liaison officer.

Planes flown: T-37, T-38, F-16 (various models/blocks).

On May 1, 1984 over the Great Salt Lake desert northwest of Salt Lake City, Utah, Meyer was part of a F-16 mission, 421st T.F.S., conducting training for new pilots. It was an interdiction mission. His plane caught on fire and started a right roll at about 3,000 ft. which could not be corrected as the flight controls had burned through. Meyer called his position to a local radar facility and pulled the handle between his legs. He said time really slowed and he remembers to this day what every gauge read and also waiting forever for the seat to give him the kick (actual time less than a second).

Once he got the jolt, he saw the F-16 fall away, do a split-s, hit the ground at an 89 degree angle and disintegrate. He had lost his sight as the wind blast hit him (ejected at about 450 knots) but did feel the wind blast, and the seat automatically separating from him. The next thing he remembers, was waking up in the chute. He accomplished a four-line jettison, allowing more steerability, and enjoyed the ride down. Meyer hit hard and landed on a cactus. He had substantial bruises and an injured neck. A U.S.A.F. helicopter picked him up about 20 minutes later and brought him home to a warm welcome.

Currently active duty major.

Awards: Honor graduate Joint Firepower Command and Control Course - 1989; 56th Tactical Training Wing/DO rated Junior Officer of the Year - 1986; 388th Tactical Fighter Wing Top Gun RLADD - 1984; Air Training Command Flying Training Award, UPT class 82-04; Distinguished Graduate, UPT class 82-04; Williams A.F.B. Support Junior Officer of the Year - 1980; Distinguished Graduate, Aircraft Maintenance Officer Course - 1979; Meritorious Service Medal; Air Force Commendation Medal with two OLC; Air Force Outstanding Unit Award; Combat Readiness Medal; Air Force Overseas Ribbon; Air Force Longevity Service Award (one Device); Small Arms Expert Marksmanship Ribbon; Air Force Training Ribbon.

Married to Barbara Ann Hohensee Meyer Dec. 5, 1986. Two boys.

Earned B.S.-biological science/pre-med (military-U.S.A.F. Academy Colorado 1979), master of aeronautical science - Embry Riddle Aeronautical University, 1988.

Active in scouting, church, raising two boys, fishing, hunting and camping. Resides in Hinesville, GA (military assignment).

**CHARLES R. MILLER,** was born May 1, 1924 in Elmira, NY. He dropped out of high school to serve in the Army Air Corps during World War II. Later he also served in the U.S. Air Force Reserve post-war. Joined the service Dec. 7, 1942, the first anniversary of Pearl Harbor. An opportunity came along for those having a high school diploma to take entrance exams for Aviation Cadets. Miller knew they were anxious for cadets and probably wouldn't really check his records. He indicated he had graduated and was allowed to take the exams. He passed the tests but failed the physical examination. He was put in navigation school and commissioned.

Military locations: Selma, LA and Alexandria, LA - navigator training; Foggia, Italy - combat; Stalag Luft I - Barth, Germany (POW). Reserve assignments: Hancock Field, Syracuse, NY; Stewart AFB, Newburgh, NY; Loring AFB, Limestone, MN; Sampson AFB, Geneva, NY.

During World War II, Miller was a navigator on the B-17 Flying Fortress. In the Reserves he was an air police officer. He served with the 15th A.F., 301st B.G., 419th B.S.

Shot down twice. Made it back to friendly territory the first time. Shot down the second time on Valentines Day, Feb. 14, 1944 at Verona, Italy on his 13th mission. He was captured by the Italians (beat him) and was literally rescued by the Germans and taken to Luftwaffe hospital. Liberated by the Russians on his 21st birthday, May 1, 1924. Arrived back home on Mother's Day, June 1945.

Achieved rank of major. Discharged Nov. 20, 1971. Retired Reserves.

Married high school sweetheart June

29, 1946. They have a son and a daughter and four grandchildren. Resides in Fort Myers, FL.

Miller returned to school and graduated from Hobart College in Geneva, NY. He then got his masters degree and received a fellowship for post-graduate work. He went on to become a high school teacher, guidance counselor and principal and always had empathy and understanding for high school and college students who seemed to lack motivation. His professional commitment was to always assure that any student wanting that second chance got it. He was later superintendent of schools, a college dean and assistant to the college president. He retired in November 1980 and is active in several community activities and organizations.

**DONNELL F. MILLER,** T/Sgt., became a member of a combat crew at Biggs Field, El Paso, TX and a member of the 493rd B.G., 861st B.S. at McCook, NE and transferred to Debach, England in May 1944.

He started his combat missions on D-Day, and his B-24 was brought down by heavy anti-aircraft fire on June 14, 1944.

Miller bailed out over the English Channel and the wind blew him into France where he was captured by the Wehrmacht. In due time he was sent to Stalag Luft IV and remained there as a POW until Feb. 6, 1945 when Stalag IV was evacuated. He was force marched cross country, which was commonly known as the Death March. He was finally liberated on May 2, 1945.

Awards: World War II Victory Medal, EAME, Purple Heart, POW Medal.

Miller married and had two children. He spent the next 40 years in the construction business. He is now retired and lives in Castro Valley, CA.

**EDWARD C. MILLER,** was born April 25, 1920 in Brooklyn, NY. Entered active duty May 2, 1942. Took pilot training in the Southeast training command and received his pilot's wings June 30, 1943.

Flew the following types aircraft: P-40, P-47, B-25, B-26, A-26, B-17, B-24, and C-47.

While serving in the 8th A.F., 93rd BG, 328th Bomb Sqdn., his B-24 was shot down

by flak on a mission to Ludwigshafen, Germany on Jan. 7, 1944. Evaded through Germany and France and crossed the Pyrenees into Spain. Finally flew from Gibraltar back to England and returned to base on June 1944.

Released from active duty Jan. 26, 194 as captain.

Worked for the Spury Corp. in New York in quality engineering for 43 years before retiring. Reside in Sedona, AZ.

**GLENN E. MILLER,** was born on Feb. 8 1924 in Flintville, PA. He joined the service on Oct. 23, 1945, serving in the Army Air Force in Keesler, Detroit, Colorado Springs, Mount. Home, Spokane, Kurmitola, India. Discharged Jan. 5, 1946.

Became a Caterpillar March 12, 1945 Mission was to deliver gas to China. The aircraft had experience some trouble with the carburetor earlier but everything seemed OK. Met with disaster when they hit a storm of wind, rain and snow. The compass wouldn't work because of the electrical storm and they got lost over India. When the gas ran out the order was given to bail out. Miller was to go first. They were at about 19,500 ft and he was a little groggy from lack of oxygen. He threw his flashlight out and jumped after it. He was afraid the chute would collapse in the storm and waited to pull the ripcord. Miller was falling through snow, wind and rain at about 100 mph when the chute opened and a terrific jar pushed his stomach up and the leg straps were tight around his legs. He landed in a plowed field 25 minutes after leaving the plane.

Awarded: Asiatic Pacific Service Medal with three Bronze Stars, Air Medal, Good Conduct Medal, Distinguished Unit Badge,

American Campaign Medal and WWII Victory Medal.

He and his wife Betty have one child and three grandchildren. Resides in Myerstown, PA, where he is mayor.

**JOHN A. MONTORO,** was born on March 12, 1919 in Hakalau, HI. Joined the service Oct. 21, 1941.

Training: gunnery school - Seven-mile Gunnery Camp, Spokane, WA. From gunnery school, went to Columbia, SC. From South Carolina went to sub-action coastal patrol in Miami, FL and Cuba. Then went to Greenville, SC and was chief gunnery instructor for two years. Trained in B-25s as upper turret gunner.

Served with the 17th B.G., 324th B.G., 321st B.G., 448th B.S. Discharged Oct. 6, 1945 with rank of staff sergeant.

Volunteered for overseas duty and flew southern route to North Africa and Corsica in April 1944 as upper turret gunner.

Shot down, July 26, 1944, on 28th mission over Verona, Italy and had to bail out of ship when the left engine was blown away. Three of seven crew members got out. Was taken into Germany as POW for nine months in Stalag Luft III. Shrapnel in leg. Liberated by General Patton's 3rd Army in 1945.

Awarded: EAME Theatre Medal, American Theatre Medal, Air Medal with two OLC, Good Conduct Medal, American Defense Service Medal, Purple Heart and POW Medal.

Married to Rose and has three children: Dan, Penny, Guy. Enjoys golf, woodworking hobbies and some traveling.

**ERNEST T. MORIARTY,** was born Feb. 21, 1922 in Winchendon, MA. Graduated Murdock High, Winchendon in June 1939.

Enlisted U.S. Army Air Corps June 26, 1941, Boston, MA. Basic training 34th BG,

18th Bomb Sqdn., Westover Field, Chicopee, MA. Group to Pendleton, OR after war started.

Transferred to Perry Institute, Yakima, WA. Graduated as engine mechanic May 27, 1942. Assigned to 306th BG, 368th Sqdn. at Wendover Field, UT.

Overseas on the *Queen Elizabeth*, reaching Thurleigh, England on Sept. 6, 1942.

Volunteered for gunner course at Bovington, England. Flew first mission on Jan. 3, 1943 to St. Nazaire, France. Flew first German mission to Wilhelmshaven Jan. 27, 1943.

Shot down in occupied France on March 8, 1943 on ninth mission. Otto Buddenbaum, pilot K.I.A. Some on crew badly wounded. All P.O.W.s. Evaded and escaped on fishing boat with 16 Frenchmen and one French woman. Landed in England April 1, 1943.

Returned by Pentagon to States for lecture tour. Returned to England July 13, 1944. Assigned to 96th BG, 337th Sqdn. Crashed on railroad tracks on 19th mission on Nov. 21, 1944. Crew in hospital. Finished last six to complete tour of 25 as spare engineer. Made a round-robin to Russio, Italy and then returned to England.

Returned to States Feb. 1, 1945. Discharged June 6, 1945.

Married second time with two daughters. Son and daughter from first marriage and three grandsons. Daughter died at age 18.

Author "One Day Into Twenty-Three", self-published with wife, Maggie.

**JOHN P. MULVIHILL, JR.,** was born Feb. 27, 1920 in Long Branch, NJ. Joined the service May 3, 1943. Served with the 15th A.F., 485th B.G., 828th B.S., 55th Wing in Venosa, Italy. Flew B-24 Liberator.

On Feb. 19, 1945, on his seventh mission, going to Graz, target area, an engine was lost. Returning two more were lost and they bailed out at 8,700 ft. over Yugoslavian coast. Mulvihill counted seven chutes. He landed on a small rocky terrain uninhabited island. Both ankles were broken and he had many other cuts and wounds. He suffered through the night and the next day two Yugoslavians came and helped him the best they could. Eventually he was turned over to the British Air Sea Rescue. He is considered an evadee.

The crew navigator was the only other survivor. Mulvihill was the bombardier.

Awards: Air Medal, Purple Heart, American Theatre, and European Theatre Medals.

Married Peggy (deceased) Nov. 24, 1945 and has one child.

Earned B.S. degree in finance in 1943 from Rider College.

Self-employed realtor. Semi-retired in 1982. Resides in Fair Haven, NJ.

**JOHN A. MURPHY (JACK),** was born Jan. 18, 1916 in Grand Junction, CO. Joined U.S.A.F. in September 1942. Earned A.B. degree.

Served with the 8th A.F., 306th B.G., 423rd B.S. Various stateside locations as well as Germany.

Became a Caterpillar following mid-air collision over England returning from mission on Dec. 15, 1944. One of two survivors.

Most memorable experiences: mid-air collision, Berlin Air Lift, and surviving 27 missions over Germany.

Divorced. Has four children, nine grandchildren and four great-grandchildren.

Worked for Continental Can Company as purchasing agent.

Discharged Oct. 4, 1969 with rank of colonel.

Resides in Huntington Beach, CA.

**JAMES G. MUSAT,** (code name Skinny) was born June 30, 1920 in Cleveland, OH. Joined the service November 1940. Federalized with 112th OBS. Sqd. D.N.G. Went to flying school and eventually joined the 422nd F.S., 9th A.F., E.T.O. then 155th Night Photo Sqd., 9th A.F., E.T.O. Discharged in October 1945. Recalled in October 1950 for Korean War and served until November 1952. Served the second time, stateside, assigned to the training command at Biloxi, MS, Keesler Field and Waco, TX, James Connolly AFB as an air and ground instructor for all-weather fighter cadets.

Planes flown: started out in 1937 with Ohio National Guard 0-47 and 0-38 open cockpit fabric covered Bi-plane, two-seater - flying school - PT-15, BT-13, AT-6 - RP322 (P-38), A-20, P-61, Black Widow Northrup, (1950 Korean recall) mostly B-25 then all-weather jets F-94 Lockheed and F-89 Northrup.

Married 37 1/2 years to Mary. Married Norma Sept. 12, 1987. Has two sons and one grandson by first wife and two daughters, one son and three grandsons with Norma.

Graduated from high school in 1939, attended Western Res. University 1945-50 nights.

Was a buyer for the city of Cleveland, metallurgist for Forest City Foundry and clerk and manager in sales department of Joseph and Feiss, Company. Also was councilman - city Fairview Park, OH. Retired in April 1983.

Discharged first time in October 1945 and second time in November 1952 with rank of first lieutenant.

Resides in North Olmsted, OH. Enjoys bowling, fishing, and travel (went to Australia four times in last six years).

**PAUL D. MYERS,** was born Aug. 26, 1919 in Volga, IA. Enlisted in Air Corp Oct. 19, 1942.

Training: radio schools at Sioux Falls, SD; Philco Radio, Philadelphia, PA, and Trauz Field, Madison WI. B-29 training at Walker AFB. Victoria, Kansas on Al Abranovic crew aboard the "Ramp Tramp."

Flew from Walker Field April 13, 1944 to Piaradoba, India and Chentu, China. Landed West Field Mariana Islands April 29, 1945 and left August 7, 1945 after completing 35 missions.

Served with the 58th Bomb Wing, 462nd B.G., 770th B.S. flying B-17's and B-29's.

Most unusual mission: took off at 20:48 for Nagoya with #3 torching. Extinguishers useless so crew successfully bailed out. Myers came down in the ocean approximately six miles east of Tinian and swam ashore about an hour later. Their B-29, Phony Express, did not fare so well as it crashed and burned near North Field.

Discharged Nov. 1, 1945 with rank of staff sergeant.

Awards: Distinguished Flying Cross, Air Medal with two OLC, Distinguished Unit Citation, Pacific Theatre Ribbon with five Bronze Stars.

Owned and operated retail jewelry store in West Palm Beach, FL from 1950 to retirement Dec. 25, 1979.

Married Beverly (deceased) Sept. 16,

1950. They have two children and three grandchildren.

Myers enjoys HAM radio and music. Resides in Murphy, NC.

**JAMES A. MYL,** Capt., U.S.A.F. (Ret.), was born Oct. 9, 1922 in Sewickley, PA. Enlisted June 20, 1942. A graduate of pilot class 43-J, Marfa, TX, from June 15 to August 25, 1944 (70 days), Myl flew a full tour of missions as B-17 pilot with the 8th A.F., 351st B.G., 511th B.S. (Polebrook, England). Also served with sixth Ferrying Group, 5th OTU, 553rd AAF Base Unit, ATC; Hq. 5th A.F., Nagoya, Japan.

On Aug. 9, 1944 (Munich, Germany - 29th mission), the Group was recalled. Near Antwerp, Belgium, the nine-ship formation in which Myl led the low squadron, was hit by a textbook flak attack. A plane to his right blew up and flamed down. Later, when Myl's plane was over the North Sea about 80 miles from the nearest land, the plane's right wing erupted in flames. Myl immediately ordered bailout, knowing there was no time to call for help. Nine airmen hit the silk in less than 10 seconds, just before the plane blew up.

The Bjoring-Martin crew saw the bail out and left the formation to circle over the array of descending parachutes. A long Mustang pilot from Steeple Morden also saw the parachutes and the splash downs and circled overhead for more than an hour to call for help. Two Thunderbirds of Air-Sea Rescue arrived much later to drop smoke flares that helped an Air-Sea Rescue launch locate the airmen and take them aboard.

The survivors were put ashore at Felixtowe and hospitalized overnight for exposure. In 56 degree water, seven of nine airmen, by using only Mae West life preservers, had stayed alive in the North Sea for over three hours. The unstoppable chatter of his teeth caused Myl to lose a filling.

Awards: Distinguished Flying Cross with OLC, Air Medal with three OLC, Presidential Unit Citation with two OLC, General Citation, E.A.M.E. Service Medal with four Battle Stars, American Theatre Service Medal, Army of Occupation Medal - Japan, WWII Victory Medal, Good Conduct Medal, Caterpillar and Goldfish Awards.

Discharged Oct. 20, 1947 with rank of captain.

On Sept. 4, 1948, Myl married Dolores E.

DuBay. They have six children and seve grandchildren. While he was self-employe in the insurance business, the Myls began purchase commercial real estate properti which two of their sons now operate. M retired at age 55 at Los Alamitos, CA, whe he and Dolores live.

Wrote 11 books focused mostly on fan ily which contain some four millions word He is St. Hedwig Church Finance Committe member; Los Alamitos-Rossmoor Libra Program chair; Polytechnic High School 50 reunion committee member for the class 1940; and 351st Bomb Group Associatio public relations specialist.

"S-A-L-U-T-E".

**WILLIAM NEWBOLD,** was born Jun 16, 1920 in Langhorne, PA. Joined U.S.A. June 16, 1941. Service in Shipham, East Angli England, Tunis, Benghazi, North Afric Served with 8th A.F., 44th B.G., 67th an 506th squadrons.

He remembers the low level Ploesti Rai Aug. 1, 1943 and the Kiel, Germany raid Ma 14, 1943.

His most outstanding mission: 67t Sqdn., Vienna Neustadt Raid Oct. 1, 194 Coming out of Africa, "on our left turn at II 14,000 feet, German fighters struck (five c six in formation, nose attack, out of the sur the ship rocked from stem to stern. We veere sharply to the right and down. The righ wing and No. 4 engine were on fire. Onl three of 10 crewmembers got out."

Newbold was discharged Jan. 13, 194 with rank of captain. Awards include th Purple Heart, Air Medal with five Oak Lea Clusters, Distinguished Flying Cross, Pris oner of War Medal, and the Presidentia Group Citation.

He has a B.S. degree in engineering Retired January 1982 as engineer in spac program. Married to Elizabeth and has thre children and four grandchildren. They re side in Langhorne, PA.

**BENJAMIN L. O'DELL,** Capt., was bor Aug. 30, 1921 in Cumberland Gap, TN. Fron Reserve, called to active duty April 194 Basic training Keesler Field, MS, college train ing detachment U.T. Knoxville, TN, primar and advanced navigation training Selma Field, LA, combat training Pyote, TX.

Flew with 359th Sqdn., 303rd BG, 8th A.F., Molesworth, England. Navigated 8th A.F. lead plane (B-17) over target at Merseberg, Germany on 30th mission Jan. 10, 1945. Navigator on lead plane to Cologne, Germany, took a direct hit from flak. Bailed out in Belgium, Battle of Bulge area. Aided by F.F.I.

Returned to States March 1945. Released from active duty to reserve September 1945.

Graduated University of Tennessee 1950. Recalled to active duty for Korean War January 1951. Served at Kelly Field, TX.

On C-97 set world record time non-stop Hickam Field, HI to Kelly Field, TX April 1951. Served as group navigator, 374th Troop Carrier Group at Tachikawa, Japan. Also manpower management officer at 374th Troop Carrier Wing. Released from active duty September 1952.

Married Ruth Rogers (deceased) and has a daughter, Linda Pierce. Grandchildren: Jennifer, Heidi Lynn, Jerry Michael. Retired executive, dairy industry. Resides in Johnson City, TN.

**RALPH E. PAGE,** was born in Chevy Chase, MD. Joined the Army Jan. 4, 1938. Served in England with the 8th A.F., 303rd B.G., 427th B.S.

Navigator on B-17 and several other aircraft.

On Aug. 15, 1944, on his 17th mission, the plane was set afire by fighters (FW-190s) after turning away from target (Weisbaden). Bailed out at 25,000 ft. and was taken prisoner on touching down. Three were killed and six were wounded on ship. Prisoner at Stalag Luft III.

Discharged Nov. 17, 1945 with rank of first lieutenant. Awarded POW Medal.

Married to Joyce E. Page. They have four sons and one grandson.

Retired C & P Telco. Resides in Rockville, MD. Enjoys golf and walking, etc.

**RALPH K. PATTON,** was born Aug. 16, 1920 in Wilkinsburg, PA. Enlisted in the U.S. Army Air Corps April 17, 1942. Received pilots wings and commission as a second lieutenant at Altus, OK, class 43E, May 23, 1943.

First duty station, Pyote, TX, co-pilot B-17 June 1943. Assigned to the 331st Sqdn.,

94th BG, 8th A.F. at Bury Saint Edmonds, England, Oct. 21, 1943. Flew first combat mission to the Renault factory near Paris Nov. 26, 1943.

Shot down on ninth combat mission to Merignac Air Field east of Bordeaux. Parachuted from crippled B-17 in the center of the Brittany Peninsula of western France at noon Jan. 5, 1944.

Lived with various members of the French Underground in the villages of Plouray, Langonnet, and Guingamp until March 18, 1944. Departed the French coast the night of March 18th from the town of Plouha via British motor gunboat 503 under the control of British Military Intelligence Service's Reseau Shelburn. Reseau Shelburn was under the command of two Canadian Intelligence officers, Lucien Dumais and Raymond LaBrosse.

Returned to the U.S. April 20, 1944 and served until the war's end as an instructor pilot B-17 at Columbus, OH and Sebring, FL.

Married to the former Bette Lou Hopkins May 1, 1944. Two children: Geoffrey L. Patton of Washington, D.C. and Beverly Patton Wand of Madison, NJ. Two grandchildren: Christopher and Elizabeth Wand.

Returned to civilian life in October 1946 as manager, Order Dept., Consolidation Coal Co., Pittsburgh, PA. Served in various managerial positions in Buffalo, NY, Rochester, NY and Detroit, MI. Retired in 1983 as vice president of eastern sales of Consolidation Coal Co., a subsidiary of the Dupont Co.

Founded the Air Forces Escape & Evasion Society in 1964 and served as its president from that time until the present. Served eight years as a director of the 8th Air Force Historical Society, including one year as president. Currently a director of the 8th Air Force Memorial Museum Foundation.

**LOUIS J. PERNICKA,** was born Dec. 9, 1923 in Chicago, IL. Enlisted in the Army Aviation Cadet Program in August 1942.

His training began at San Antonio Aviation Cadet Center and ended four-and-a-half years later at the same base as squadron commander of base maintenance operations.

His stateside training in fighter aircraft was all in P-40s. Assigned to 57th F.G., 66th F.S. which was flying the P-47 Thunderbolt.

The 57th's prime objective was dive bombing and strafing bridges, railroad yards, airfields, flak gun emplacements, convoys, close infantry support, etc.

The objective of his first combat mission was to divebomb a bridge and strafe the area but his P-47 was hit by flak on the initial run and severely damaged the flight controls. Force to bail out over the Tyrrhenian Sea and was picked up by the British Air Sea Rescue Service, floating in a dinghy.

During the next 80 missions he wiped out three more P-47s; chased an Me-262 but couldn't close within firing range; flew through high tension lines and brought back 1/4 mile of wire; downed an Me109 (just the fuselage off a rail flat-car); in addition to accomplishing the objectives of the 57th.

On his 82nd mission, after leading 12 P-47s in divebombing a railroad bridge, he spotted a motor transport and began a series of strafing runs. A 40 mm shell met him in the cockpit, shattering the bones in his left shoulder, breaking a rib and peppering his face, arms and legs with fragments as well as "that part which distinguishes man from woman". He was able to belly-it-up on a friendly strip. After seven months in the hospital he returned to flying status.

Pernicka married "Sweet Violet" for which the old "97" was named. They have three children, Gayle and twins Robert and Susan and four grandchildren.

He earned a B.S. in engineering from Illinois Institute of Technology and did graduate work at Columbia University.

Discharged January 1947 with rank of captain. Resides in Norwalk, CT.

**NICHOLAS J. PETERS,** was born Nov. 15, 1920 in Wyandotte, MI. Joined the service in March 1942 Ft. Custer, Battle Creek, MI. Basic training (Infantry) at Camp Livingston; 112th Infantry, H Co., 28th Div. Alexandria, LA. Transferred to Army Air Force, San Antonio, TX, Nov. 18, 1942 and went to Aviation Cadet Center for pre-flight (pilot) training Jan. 19, 1943-March 20, 1943. Elementary flying at Cimarron Field, Yukon, OK, 310 A.A.F.C.F.S., March 21, 1943-June 25, 1943 and basic flying June 27, 1943-Sept. 28, 1943 at Strother Army Air Field, Winifield, KS.

Other schools and training at: Childress

Army Air Field (bombardier), Childress, TX; Lowry Army Field (armament school), Denver, CO; Buckingham A.A.F. (flexible gunnery school 23 A.T.U.) Fort Meyers, FL; Drew Army Air Field (327 A.A.F. Base Unit R.T.U. HB) Tampa, FL; Hunter A.A.F. (3 A.A.F. staging wing) Savannah, GA; Goose A.A.F., Goose Bay, Labrador; Goodfellow Air Force Base (A.T.C.) San Angelo, TX; Webb A.F.B., (missile training wing) Orlando, FL. Stationed at numerous other and various locations stateside, in Europe, the Pacific and Asia.

Shot down on 11th mission Jan. 5, 1945 over Germany. Completed 35 missions.

Achieve rank of staff sergeant. Discharged in July 1945. Reenlisted. Discharged in July 1965.

Awarded Air Medal with Silver Oak, WWII Victory Medal, E.T.O. with four Bronze Stars, Army Good Conduct Medal, Distinguished Unit Badge with Bronze Oak, Air Force Longevity Service Ribbon with four, Air Force Good Conduct with three Bar Loops, Navy Good Conduct with one star, Navy Expert Pistol Medal, American Campaign Medal and National Defense Medal.

Retired from Air Force and Post Office.

Married to Jean for over 41 years. They have three children and three grandchildren.

Worked for the U.S. Postal Service. Retired Oct. 31, 1986.

Resides in Tamarac, FL.

**ROY A. PETERSON,** 1st Lt. U.S.A.A.F., was born June 23, 1922 in Flint, MI. Enlisted Aviation Cadets Aug. 16, 1941. Served as pilot with 8th A.F., 389th B.G., 567th B.S. (B-24) Flew over 24 combat missions.

Final mission was to Kassel, Germany Sept. 28, 1944. As formation was nearing Koblenz, flak caused a fire in the number four engine. Another burst tore off the right wing. Plane fell into a tumbling spin and bailout was made at about 25,000 ft.

Peterson hit something as he was exiting the top hatch and fell unconscious about four miles before opening his chute. Landed in trees in the Hunsruck. Received facial fractures and lacerations. Captured along the Moselle River after evading for five days.

Awarded Air Medal with OLC and POW Medal.

Retired from Buick Motor Division Manufacturing engineering in 1984.

Married to Dona Coe and had nine children. Resides in Flint, MI.

**JOHN W. PETTY (BILL),** was born March 22, 1918. Joined Air Force Nov. 11, 1942. Served 1943, Cochran Field, GA on basic training planes as mechanic. Gunnery school, 1944 at Harlingen, TX. Overseas combat training on B-24 bomber in Casper, WY.

Nose gunner, B-24 *Dinah Might*, 761st Sqdn., 460th BG, 15th A.F., Spinazzola, Italy, Lt. Gerald S. Armstrong, pilot.

After his bomber crashed in enemy territory, Petty evaded capture and was rescued by two men (still alive) in Yugoslavia. Two other crew members and he were hidden by a lady for five days and nights. They walked through enemy territory for 72 days, two-to-three hundred miles. Later flown out of Yugo. back to friendly territory in south Italy.

In 1969, while in Venice, Italy, he wanted very much to find the lady, Mrs. Amaleja Faletic, to thank her for saving him from capture by the enemy and possibly from freezing in the Alps. What once was Caporetto, Italy was now Kobarid, Yugoslavia, the boundary being changed after the war. He found the lovely 88-year-old lady on a return trip to Yugoslavia in 1969.

Currently resides in Carthage, TN.

**PAUL W. PIFER, M.D.,** was born Dec. 2, 1924 in Punxsutawney, PA. Entered U.S.A.A.F. Dec. 7, 1942. Served with the 446th B.G., 2nd Air Division, 8th A.F.

Was a radio operator-gunner, one of ten men in a B-24 bomber shot down by the Germans over Trier Germany, Jan. 1, 1945. Four of the parachute did not open probably due to the extremely high altitude and the airmen had to open them by hand. Pifer and two others were successful pulling out the chutes from the front of the packs with their hands. Another was not. Pifer suffered a broken pelvic bone in the bailout and was sent to a hospital in France near where friendly Partisans had helped the flyers who had landed in hostile French territory.

Flew 30 missions over Germany. Shot down a second time but was able to make an

emergency landing at a fighter base in France. The third time he was shot down, this time 600 miles off the coast of Spain, the crew was able to dump their excess weight and make an emergency landing in a sheep pasture.

Married Ruth and they have two children. Member of F.A.C.S., F.A.C.O.S. National surgeon of the Caterpillar Association.

Semi-retired. Resides in Covington, LA.

**IRWIN J. PIRE, JR., (SAM),** was born March 3, 1924 in Madison, WI. Entered U.S. Air Force Oct. 30, 1942. Served with the 8th A.F., 384th B.G., 544th B.S. in England.

Shot down on seventh mission, Nov. 8 1944, over Frankfurt, Germany. Lost all but one engine. Plane was on fire. Bailed out at low altitude (less than 1,000 ft.). Saw bombardier hit the ground dead (his chute did not open). Nine of the original crew are still alive.

Pire hit the ground hard and was knocked out. He was dragged through vineyard and barbed wire. Had a painful dislocated shoulder and injured back, knee and ankles and had many cuts and wounds. He tried to escape but was too injured. He had been captured and taken POW. On the 600 mile Death March.

Discharged Oct. 30, 1945 with rank of technical sergeant.

Awarded Purple Heart and Service Connected Medals.

Retired as plant manager May 1, 1986. Resides in Strum, WI.

Married Norma C. Pire and they have four children and eight grandchildren.

**ORVIS C. PRESTON,** was born July 2, 1923 in Nowata, OK. Entered the service Dec. 8, 1942. Served with the 3rd and 9th A.F., 410th B.G. (L), 646th B.S. Trained (armory) at Denver, CO, gunner at Ft. Myers, FL, other training at Muskogee, OK.

Planes flown: A-20, B-25 and B-26.

Discharged Oct. 8, 1945 with rank of staff sergeant.

Married Lorraine M. Feb. 1, 1946. They have three children and four grandchildren.

Went to college for two years. Retired after 50 years as a plumber.

Currently involved as commander, Fox

Rider Valley Chapter of American Ex-Prisoners of War.

Resides in Northlake, IL.

## FRANCIS C. RAMSEY (FRANK), was

born Feb. 11, 1921 in Gaffney, SC. Joined the Army Air Corps Aviation Cadet Program June 18, 1942.

Took basic training at Keesler Field in Biloxi, MS; aerial gunnery training at Laredo, TX; and airplane mechanics at Keesler Field.

Assigned out of Salt Lake City, Utah as a member of the original 464th B.G. (H). Took combat training at Pocatello, ID. Flew overseas by the southern route. Flew first mission out of Gioia, Italy and later moved to permanent base at Panatella, Italy. Served with the 776th B.S., 464th B.G. (H), 15th A.F.

On his third trip to Ploesti, Rumania, his 39th mission, his B-24 developed engine trouble on two engines. They increased manifold pressure on the other two engines, stayed in formation long enough to drop their bombs on the target but were unable to maintain altitude. The pilot, Captain Bruce C. Cater gave the order to bail out over Yugoslavia.

Ramsey was able to unite with four members of the ten man crew. They had bailed out in a rugged section of the Kapela Mountain range. Spent 39 days there as evader. Helped by Partisans and eventually flown out by the Russians.

Stateside he trained at the engine specialist school at Chanute Field, IL. After graduating as an engine specialist with flying status, he was sent to the B-29 base at Savannah, GA. He stayed there until discharge Aug. 29, 1945. Achieved the rank of technical sergeant.

Awarded: ETO Ribbon with seven Battle Stars, two Presidential Unit Citations, the Purple Heart, Air Medal with two OLC, and the Distinguished Flying Cross.

He graduated from Clemson University in 1948 with a B.S. degree in textile engineering.

Married Hazel M. Spencer in 1949 and they have two children, Steve and Janice.

Retired from the textile business in 1980 and from golf the same year. Enjoys travel and reading. Resides in Gaffney, SC.

## JOHN F. RAUTH, (STUBBY), was born

Jan. 15, 1921 in Wabasm, NE. Joined the service Aug. 14, 1940. Was with second A.A.C.S. - pilot school - 319th F.S., 325th F.G., advance pilot instructor, 139 MAG. Stationed at Mitchell Field, NY, Africa, Italy, Craig Field, AL, Rosetrms Mem. Airport, St. Joseph, MO.

Planes flown: PT-17, PT-13, T-6, P-40, P-47, UC-78, B-26 - Douglas, T-33, F-80, RF-84F, C-97. Total flying hours: 4,000 hours.

Shot down on a Friday, July 23rd on a fighter mission over a small island off the coast of Sicily. Rauth bailed out some place under 700 ft., going over the right side and under the horizontal stabilizer. As soon as he was clear of the plane he opened his chute but before he could completely get out of it, he hit the water. The airplane had hit the water before his parachute opened. The wind carried his chute and dragged him "like a surfboard" for some distance. He spilled the chute but got tangled in the shroud lines. Stayed afloat in his Mae West, cut free of the parachute and inflated his dinghy as his flying school roommate circled above. Picked up by a fishing boat. Taken prisoner by Italian soldiers. Was fed, interrogated, and transported to various locations around Trapani during the seven hours held captive as U.S. paratroopers were taking the area. His captors became his prisoners.

Married Betty Aug. 5, 1944 and they have seven children and 11 grandchildren.

Attended college for two years. Operated own construction company 1953-81. Now works for son in same business.

Retired from the military Jan. 15, 1981 with rank of lieutenant colonel.

Currently is a college art student and kitchen planner in son's business. Resides in St. Joseph, MO.

## KENNETH LEON REED, S/Sgt., was

born Jan. 11, 1932 in Keota, OK. Joined the Army Oct. 24, 1950. After basic training served with the 101st Airborne in Korea until October 1953. Discharged from Army and re-enlisted in Air Force 1954-72.

While on classified mission transporting hazardous cargo (H.E.) aboard C-124, Globemaster, on March 8, 1957, two engines were lost and they were losing altitude quickly. Bailed out over rural Kansas farmland.

Military locations include Osan and Kimpo, Korea, Goose Bay, Labrador, Thule, Greenland, Sembach, Germany, Anderson AFB, Guam, Parks AFB, CA, and BAFB, LA. Flew C-124, KC-97, and H-19. Discharged Jan. 31, 1972 with rank of staff sergeant.

Awards from Army and Air Force: Korean Service Medal with four Bronze Stars, United Nations Service Medal, Korean Ser-

vice Medal, Good Conduct Ribbon with rope and five knots, WWII Occupation Medal of Germany.

Presently retired from farming and keeps occupied gardening and doing lawn work. Married to Mayme Reed, July 14, 1957, and they have four children and six grandchildren. Resides in Tuckerman, AR.

## EUGENE J. REMMELL, was born Aug.

8, 1928 in Baltimore, MD. Entered service Sept. 2, 1941. Sent to Sheppard Air Force Base, TX, for six months aircraft mechanics school training. Completed aircraft and gunnery training at MacDill AFB in Florida. Joined the 91st B.G. at MacDill in April 1942 and after training there and several other places, left for England in October 1942 to join the 8th A.F.

Completed 25 missions as a flight engineer, top turret, gunner, on B-17 aircraft. Was a gunnery instructor in England for several months before returning to U.S. in October 1943.

Was assigned to train B-24 gunners and volunteered for another tour of combat on B-24's. Sent to Italy with 15th A.F. in August 1944. Flew as flight engineer gunner with the 450th B.G., 720th B.S. Wounded and bailed out returning from a combat mission to Vienna, Austria on Oct. 7, 1944. Got back to Italy with the help of Marshall Tito's Partisan soldiers after walking many miles across Yugoslavia. One man was killed in the bailout, five were returned with help of Partisans, and four were POWs.

Awarded: Distinguished Flying Cross, Air Medal with nine OLC, Purple Heart, ETO Ribbon with seven Major Battle Stars and Presidential Unit Citation.

After his return to Italy, Remmell flew missions with different crews to complete his tour.

Married Ruth Osborn in April 1953 and they have three children, two born at Tripler Army Hospital in Hawaii. While in Hawaii with B-29 weather reconnaissance squadron, participated in Operation Castle at Eniwetok in 1954.

Retired May 1, 1963 with rank of master sergeant from Andrews Air Force Base, Washington, D.C. Enjoys golf. Resides in Baltimore, MD.

**HARRY REUSS,** was born March 11, 1920 in Indianapolis, IN. Entered service in April 1943. Sent to Goldsboro and Seymour Johnson for basic; Keesler for A.M.; Harlingen for gunnery; Lincoln, NE for crew formation; and Davis-Monthan for phase. Picked up new B-24J at Topeka. Port of debarkation: Grenier Field. On to Italy via Gander Lake, Azores, Marakesh, Tunis and to the 15th A.F. at Pantanella. An "eager" engineer was ready!

From the 465th B.G., 781st B.S. (H), they made three successful take-offs and two successful landings in Italy. The third landing was flunked and they went "down in flames" over Vienna Sept. 10, 1944.

After numerous harrowing stops, Reuss eventually wound up in Stalag Luft IV, which was dissolved Feb. 6, 1945 as the Russian offensive began to close in on them (10,000) "fallen airmen". They were marched around Germany for 86 days. Ruess was liberated on May 2, 1945 by the British Royale Dragons (tanks).

After treatment in three British hospitals and the American 28th General Hospital, his POW 106 pounds had been augmented. With new life in his body, he took off for Paris and London. Returned to States aboard LST 983, a very small ship in the ocean. Honorably discharged at San Antonio, November 1945.

Graduated from Butler University in 1951. Life member of American Ex-Prisoners of War, MOPH, DAV, VFW and American Legion.

Married for over 43 years to Lynn. They have two children, Rick and DeeDee and two grandchildren, Melissa and Steven. Resides in Indianapolis, IN.

**ALBERT J. RILEY,** was born April 30, 1922. Entered Air Corps Oct. 6, 1942. Served with the 8th A.F. in England. Achieved the rank of staff sergeant.

Shot down over Brunswick, Germany, May 19, 1943 on his 30th and last scheduled mission. He and another crew member had to bail out the nose wheel door as the B-24 was badly damaged and fell out of formation. The other crew members were killed in the crash witnessed as Riley floated to ground in his parachute. He was captured and taken POW for a year.

Had he completed this mission and returned to his base in England his war would have been over.

Awarded Distinguished Flying Cross, Air Medal with three OLC New York State Valor Medal, and POW Medal.

Retired from the New York telephone company after 39 years of service.

Father of seven children and 11 grandchildren. Lost his wife after 45 years of a happy married life. Resides in New Hartford, NY.

**MICHAEL S. RIZZO,** was born Nov. 8, 1910 in Kenosha, WI. Entered service Dec. 31, 1942. Served in CAC and TC at Ft. Bliss, Camp Wallace, New Orleans, New Guinea and the Philippines.

Served with AA units, Transportation Corp, as air travel officer, Headquarters AFUESPAC.

Bailed out of C-47 transport over Tai Po, China.

Discharged July 30, 1946 with rank of captain.

Graduated in 1938 with LLB degree from the University of Marquette. Self-employed attorney.

Married Angeline M. Rizzo. Resides in Kenosha, WI.

**RALPH C. ROBERTS,** was born Jan. 31, 1918 in Sharon, VT. Entered the service Oct. 14, 1942. Served with the 8th A.F., 401st B.G., 615th B.S. at Bassingbourn, England. Flew B-17.

On his third mission, Jan. 29, 1945, to Frankfurt, Germany, right after dropping their bombs, a rocket hit them on the tail. The fighters then closed in and knocked out two engines and set the plane on fire. The crew bailed out at about 2,500 ft.

Roberts delayed opening his chute until about 5,000 ft. He had an anxious moment. Finally found his rip cord on the left instead of the right (where it was suppose to be). He landed in the only tree around near a German air field.

The airman was taken prisoner and spent six months in Stalag Luft VI, six months in Stalag Luft IV, then three months on the Death March. Liberated at Bitterfield, Germany April 26, 1945.

Roberts married Lillian July 16, 194 and has one daughter, Ann, and one grand son, Christopher Dame.

He was a partner in a welding suppl business. Retired Nov. 2, 1988. Now enjoy golf and bowling. Resides in Wilder, VT.

**ROBERT C. ROGERS (BOB),** Sgt., wa born Dec. 21, 1924 in Urbana, IA. Graduate Urbana High School 1942. Enlisted in U.S Marine Corps July 5, 1943.

After boot camp in San Diego, CA served in various training and fighter squad rons stateside. Embarked for overseas and joined Marine Night Fighter Sqd. VMF(N) 534 on island of Guam in the Pacific theater Returned to U.S. by way of Japan in November 1945. Served U.S.M.C. air station until honorably discharged on June 15, 1946.

Trained as aircraft mechanic. Served as crew chief on flying duties in F-4U Corsair F-6F Hellcat, and F-7F Tigercat.

Made two emergency parachute jumps from disabled aircraft during World War II The first jump was on a mission to fly a GH-3 Howard from the M.C.A.S. at Parris Island to the Naval Air Station at Atlanta, GA to escape a hurricane that was heading towards the coast. The aircraft pressure fell and the engine quit, (later findings were a broken fuel line that allowed all the fuel to escape from the tanks so fast nothing could be done). The pilot ordered bailout but the door was stuck. They eventually got the door open and Rogers was the third man to bail out. Back at the M.C.A.S., P.I. they had been given up for dead.

Rogers second jump happened in 1945 when he was stationed at Guam. He was ordered to be on a flight to Okinawa, a long flight over the ocean at night. The night was clear. Suddenly there was a loud bang and a tongue of red flame streamed back along the fuselage from the engine with 30 ft. of fire. It was dark and impossible to see the water below. Rogers landed in the water, got out of his chute and harness and inflated his Mae West. He had brought along a pack containing a life raft and inflated it and awaited daylight.

Sun up there was no sign of other men, just water. Finally rescued by air-sea rescue plane. Two other crewmen had been picked up already. Rogers was never questioned about the incident and was told they had been shot down (something he did not believe).

Awarded: U.S.M.C. Good Conduct Medal, Navy Commendation Ribbon, Presidential Unit Citation Ribbon with one Battle Star, American Campaign Medal, Asiatic-Pacific Campaign Medal with one Battle Star and the WWII Victory Medal.

Married to Christine. They have three children and two grandchildren and reside in Stone Mountain, GA. Retired as senior safety engineer after 40 years with Georgia Power Company.

**LEONARD ROSE (GENE-ROSIE),** was born in Vincennes, IN. Inducted Feb. 15, 1943 into U.S.A.A.F. Basic training at St. Petersburg, FL; radio school at Scott Field, IL; armor school at Denver, CO; gunnery school at Laredo, TX; crew formation at Salt Lake City, UT; crew training B-24s at Davis-Monthan at Tucson, AZ; crew assignment at Lincoln, NE. Troop Ship USS General Butner, Norfolk, VA, arrived in Oran North Africa. Then up the Mediterranean Sea on a limey ship, Semarie of Liverpool, to Naples to base at Cerignola, Italy.

Served with the 15th A.F., 459th B.G., 758th B.S. Shot down on 28th mission, Aug. 29, 1944. Bailed out over Yugoslavia. Taken in 40&8 railroad box car through Czechoslovakia and Poland to POW camp, Stalag Luft IV in the town of Grosstychow and the railroad station of Kiefhiede, crossing of the 16 and 54 meridian on the globe, in the Pomeranian sector of northern Germany, now part of Poland.

Walked out of POW camp by the Germans on Feb. 8, 1945. Walked until April 24, 1945, when he was liberated by the Russians who in turn locked him up for three weeks. Rose escaped and got to the American Army May 20, 1945.

He is now retired and resides in Indianapolis, IN. Married Ella and they have two children, Gina and Barry and two grandchildren, Kenny and Maddie.

In 1989 he received the highest award that the Governor of Indiana can honor a citizen with, the Council of the Sagamore of the Wabash. In 1990 he was awarded Honor-

ary Secretary of State and in 1992 in a national citizenship award, he was named Indiana's winner.

Rose is the past Indiana state commander of the American Ex-Prisoners of War. He is the current national director of American Ex-Prisoners of War and has a mailing list of 3,500 Ex-POWs from Stalag Luft IV out of a camp of over 10,000 Ex-POWs many years ago.

**JOHN C. RUCIGAY,** 1st Lt., was born Jan. 25, 1925. Joined the U.S.A.A.F. April 8, 1943. Completed pilot training in January 1944 in the southeast training command. Assigned as B-24 co-pilot with crew training at Davis-Monthan Field, Tucson, AZ.

With crew, picked up combat aircraft at Topeka, KS and ferried the B-24 to Italy via South America and Africa. Assigned to 778th B.S., 464th B.G., 15th A.F.

After flak damage had only one engine

operational, bailed out at 10,000 ft. with crew in occupied northern Yugoslavia on the 17th combat mission while returning from Munich, Germany, July 19, 1944. Rescued by Partisans and returned to Italy after 40 days.

Completed active military service stateside September 1945 after serving as pilot instructor in AT-6s. Flew C-46s in the Reserves.

Awarded: Distinguished Flying Cross, Air Medal with OLC, Purple Heart and Presidential Group Citation.

Graduated as an aeronautical engineer from Polytechnic Institute of Brooklyn and employed as a helicopter flight test engineer with Vertol.

Retired in 1986 after 30 years with General Electric as a flight control systems engineer and gas turbine project manager.

Married Dorothy Kaune in 1950 and have two daughters, one son, and one grandson. Resides in Ballston Lake, NY.

**CHARLES J. SALIVAR,** Lt. Col. U.S.A.F. Reserves (Ret.), was born Feb. 15, 1923 in New York, NY. Entered Army Air Force Oct. 19, 1942. Served in Panatolla, Italy with 15th A.F., 464th B.G., 779th B.S.

Bailed out Oct. 17, 1944 over Yugoslavia. Evaded with help of Tito's Partisans.

Discharged Dec. 3, 1945 with rank of first lieutenant. Entered Reserves.

Married Mary Jane Court Jan. 25, 1975. Have two children and seven grandchildren.

Civilian career in the pharmaceutical business. Retired July 1, 1986. Resides in Kirkwood, MO.

**SOLDIER EDWARD SANDERS,** a full-blooded Cherokee Indian, was born Sept. 20, 1918 in Stilwell, OK. Entered military service May 14, 1942, assigned to 8th AF Composite Command.

Overseas September 1942, stationed near Belfast, Northern Ireland until 1943, then attended gunnery and communication schools near Kings Lynn, England.

Participated in air offensive, European Campaign with B-17 Pathfinder Group and 390th BG, 571st Sqdn. as a gunner with staff sergeant rank.

Received the following awards: Distinguished Unit Citation, Air Medal with one Oak Leaf Cluster, EAME Ribbon with one Bronze Star.

Crew was shot down by fighter attack over Magdeburg, Germany, May 28, 1944. Spent eight months at Luft IV, three months at Stalag XIII D Nuremburg. In April 1945 while on forced march somewhere between Nuremburg and Moosburg, he and a fellow POW escaped.

He made contact with the American Army approximately two weeks later, alone, as he and his partner had become separated three days before. He was then flown to Paris, France hospital, where he was recuperating when the war in Europe ended.

Retired 1983 from government service after 43 years.

He and wife Catherine reside in Cherokee, NC. They have two sons, (both veterans of the Vietnam War era), two daughters and four grandchildren.

**LEONARD J. SCHALLEHN,** was born March 2, 1921 in Saratoga Springs, NY. Joined the military in fall of 1942.

Single-engine pilot training, Southeast training command. Assigned to 405th Fighter Grp., Christ Church, England.

Shot down June 16, 1944, over Mayenne, France. Following successful landing, three days and 30 miles later, arrived in Domfront. Helper was Andre Rougeyron, Resistance leader in that area of Normandy and mayor of Domfront. Rougeyron later arrested by Gestapo within Schallehn's vision. Sixty-five days later, rescued by Patton's Third Army.

Following Rougeyron's escape from Buchenwald, he and Schallehn embraced at first reunion in 1961. He remains in close contact with family.

Released from duty Jan. 1946, joined New York Telephone Co., (retiring 35 years later), and married Eunice Waite in July 1946. They have two married daughters, Cynthia and Sandra and two granddaughters, Kelly and Emily. Along with his family, golfing and skiing continue to be main interests. Resides in Saratoga Springs, NY.

## ROBERT A. SCHWARTZBURG,

was born Jan. 25, 1919 in Appleton, WI. Joined A.A.F. Oct. 26, 1942. Trained and stationed at Sioux Fall, SD; Kingman, AZ and Bury Saint Edmunds, England.

Shot down Jan. 5, 1944, Brest, Peninsula. Escaped, hidden and passed through French Underground to England.

Discharged Nov. 16, 1945 as staff sergeant.

Awarded Purple Heart and Air Medal.

Married Betty Louise Papez and has two children and one granddaughter.

Retired from automobile dealership management.

## BOB SCOTT, Col. U.S.A.F. (Ret.), served

as a fighter pilot during WWII and the Korean Conflict. During WWII he logged 54 missions in China and was credited with two aerial victories. During the Korean Conflict, as commander of the 35th F.S., he flew 117 combat missions in the F-86F.

During a "scissors" maneuver in July 1943 the empennage of his aircraft was cut by a propeller which resulted in loss of elevator control followed by a high speed dive forcing the use of a Switlick 24 ft. diameter parachute.

Upon Scott's return from Korea, he assumed command of the 510th Fighter-Bomber Squadron at Langley AFB for three years. Here he set an official U.S. transcontinental speed record in a F-84F Thunderstreak, using two KB-29s for aerial refueling.

His last Air Force assignment was a three year stint as commander of tactical air command's 832nd Air Division.

He is a veteran of 134 combat sorties over North Vietnam. As commander of the 355th Tactical Fighter Wing, based at Takhli AB, Thailand, he led his wing of F-105 Thunderchiefs in many strikes against the Hanoi-Haiphong-Thai Nguyen area. On a raid, March 26, 1967, he downed a MIG-17 in aerial combat.

Scott holds two degrees in aeronautical engineering and is a graduate of the Air Force Experimental Test Pilot School. He now owns working cattle ranches near Tehachapi, CA and Santa Rosa, NM. Until the age of 60, he was chief pilot for Antilles International Airlines.

Awards: four Silver Stars, three Legion of Merits, six Distinguished Flying Crosses, 16 Air Medals, plus 18 others.

Resides in Tehachapi, CA.

## THOMAS B. SCOTT, was born March 5,

1923 at Fort Scott, KS. Entered U.S.A.F. Aug. 24, 1942 and remained until June 1945. Served again July 5, 1947 to Sept. 30, 1965.

Took weather officer training, 1942-43 at Chanute Field; navigator training 1943-

44 at Ellington Field, TX and pilot trainin 1948-49 at Randolph/Barksdale. Rated "e pert" on 45 cal. Thompson sub-machir gun.

Served in 1944-45 with the 464th B.G 5th A.F. in Italy; 1951-52 staff weapons off cer, 18F B.G., K-46 Korea; 1957-58 with 56t W.R. Sqd., Yokota AB, Japan; 1963-64 wit 30th Weapons Sqd., Saigon, Vietnan Weather reconnaissance pilot.

Memorable experiences: first typhoo flew into was name of wife, Irene; flying ou of BW-1, Greenland at night in SA-16 a control team member supporting TAC; firs non-stop deployment to England with mid air refueling in 1955.

Jan. 8, 1945, flying as nose turret gun ner/navigator in B-24 with a Ma MacDonald as deputy group leader, num ber two engine caught on fire. Bailed ou nose hatch and landed in Mussolini gam preserve in Italy boot. Landed in stockin feet and broke a bone in left foot. Grounde for 30 days.

Later heard that pilot flew plane bac to base and landed by himself. Fire to en gine went out after gas shut-off as co-pilo was leaving aircraft. All other crew exite and landed safely (good chutes). Have pic tures of bailout taken by other aircraft.

October 1944 made an emergency land ing on Isle of Vis off Yugoslavia coast. Wais gunner deployed. Good chutes fixed to gu stanchion upon touchdown which provide braking action. Crew reported missing i action. Aircraft subsequently repaired and with added fuel, flew back to base in Italy

Married B. Irene Scott and they hav one child. Employed as director of Longview Development Inc.

Discharged Sept. 30, 1965 with rank o major. Resides in Blue Eye, MO.

## ROBERT SEIDEL, was born July 20, 192

in Euchart, IN. Entered Aviation Cade Training Program May 1943. Completed combat air crew training June 1944. Served with 15th A.F., 55th Combat Wing, 460t B.G., 763rd B.S. Air combat: Apennines Rhineland, Balkins.

On Nov. 16, 1944 hit over target, (Munich, Germany) and struggled to Salsberg, Austria. Bailed out in heavy snow storm and was separated from the rest of the crew. Captured Nov. 29, 1944. Interned at Stalag IV and I. Returned to U.S. June 1945.

Awards: WWII Victory Medal, EAME Ribbon, American Service Air Medal, Purple Heart, Presidential Unit Citation, POW Medal and Korean Conflict Medal.

Reassigned to MATS. Released from active duty to Reserves November 1946.

Earned a B.S. in mech/aero engineering. Worked on various aero space programs.

Member of AFA, Ex-POWs, VFW, CAF, AF Evasion Society, Caterpillar Assoc., B-24 Club.

Married Helen. They have three children and five grandchildren. Resides in Dallas, TX.

## DAVID SHOSS,

**DAVID SHOSS,** was born Dec. 25, 1910 in Houston, TX. Entered Air Force Feb. 6, 1941. Served with 36th CAV, Recon. Troop, 36th Div. Hq. Signal Co. in Brownwood, TX.

His ship, a B-17, received a direct flak hit forcing bailout at noon June 24, 1944 at Rouen, France. Shoss was captured. Escaped Sept. 11, 1945.

Discharged July 4, 1945 with rank of first lieutenant.

Married and has three children. Resides in Dallas, TX.

Was vice president of Zale Corporation. Retired Dec. 31, 1988. Earned B.A. degree from Rice Institute.

**HERBERT J. SKINNER,** Major U.S.A.F. (Ret.), was born Sept. 13, 1916 in Gowen, MI. Enlisted April 13, 1942.

Served in England and continent during WWII primarily with 153rd Ln. Sqd., C fight commander on detached service with Army Corps Hq. Served in numerous air bases around U.S. Had one tour in Japan, 1952-55. Retired Travis AFB July 31, 1962. Training in Gulf Coast 43-D (Brooks).

Bailed out of P-40F July 13, 1943 over Tennessee. Ran out of gas.

While attached to Third Army Corps in December 1944 during Battle of the Bulge in Bastogne, Belgium, Skinner moved his flight to the Corps Rear in France. He moved a family that was heavily involved in the Underground and British Intelligence system with him. He later married the daughter and they are still married today.

Planes flown: P-40, P-51, C-47, C-131, C-124, L-4 and L-5.

Awards: Air Medal, ETO Ribbon with six Battle Stars, United Nations Ribbon, other time and travel ribbons and the Presidential Unit Citation.

Earned B.A. degree in sociology with option in welfare and corrections from California State University in December 1968. Retired from State of California, United States Civil Service and U.S.A.F.

Discharged July 31, 1962 with rank of major.

Married to Yvonne Wauthier and they have two children and two grandchildren. Resides in Henderson, NV.

**KENNETH SLAKER,** Lt. Col. U.S.A.F. (Ret.), was born April 14, 1920 in Elderton, PA. Soloed in an Aeronca as a private pilot in November 1940. Joined the Army Air Corps Dec. 18, 1941. Received commission and pilot wings July 26, 1942.

Trained in B-17s and joined the 301st B.G. in Africa in January 1943. After completing 50 combat missions, was assigned as the base operations officer at Lincoln Army Air Base until discharged. Attended the University of Washington and received a regular commission in October 1947.

Planes flown: PT-17, BT-13, AT-9, T-6, B-17, B-24J, B-25H, A-25, C-47 and P-38.

Graduated from Air Force jet school at Randolph AFB in 1958. Flew T-33 and 100 series jets.

Assignment when retired in 1966 was as commander of the AFROTC detachment at the University of North Carolina. Awarded a masters degree in industrial management at the University of Southern California and was employed by the Boeing Company in Seattle, WA.

Fathered seven children and have been married to Lee Ann Stanislaus of Vinita, OK for over 35 years. Resides in Normandy Park, WA.

**LEWIS E. SMITH,** joined the Air Force in October 1942. After completing Airplane Engine Mechanic and Gunnery School, he was assigned to a combat training crew on a B-24 Bomber.

After completing all phases of training, he was promoted to Sgt. He and crew picked up their new B-24 and headed for the war via Southern Route into Africa and on to Italy to join the 450th Bomb Group, incidentally was the first to land in Italy December 20, 1944. February 1944, this crew became replacements as the 450th already had 90% casualties.

He flew 43 missions with the same crew: their plane was disabled on June 13, 1944, at Munich and made it back to the island of Vis in the Adriatic Sea where they bailed out. He and five others landed in the water, (four on land) the ones in water were later picked up by U.S. Air Sea Rescue. After seven days they returned to their base in Manduria, three days after that, his pilot was killed on a routine flight to 15th A.F. HQ. He flew his remaining five missions, (a total of fifty) with other crews. The last one being with Col. Giddion, 450th Bomb Group Commander.

For the record, he was over the target Ploesti five times. His parachute was manufactured by Cole of California #43-59912.

Awards include Distinguished Flying Cross, Air Medal with 5 oak leaf clusters, ETO with seven battle stars, 2 group Presidential citations. Discharged September 1945, with rank of S/Sgt.

Married to Margaret, his wife of 48 years. They have one daughter, three grandsons and one great granddaughter. He is a retired farmer.

**CLYDE W. SNODGRASS,** was born April 19, 1929 in Seattle, WA. Joined U.S. Army June 20, 1947. Regular Army appointment to U.S. Military Academy on July 1, 1948. Commissioned into U.S.A.F. June 3, 1952.

Air defense fighter pilot until 1956. Later served as Minuteman, ICBM combat crew member and operations officer/unit commander. Retired Aug. 1, 1975 with rank of lieutenant colonel.

In July 1950 was returning from Seattle area to West Point. Having "hitched" a ride by B-17 to MacDill AFB. Was en route on a C-82 from MacDill to Maxwell AFB, AL on July 31st, when an engine failed and they turned for Tallahassee. Within a few minutes the other engine sputtered and died. They were at about 1,200 ft. when directed to bail out.

The first out the door dumped his canopy on the floor. The crew kicked the pack after him and followed. He was lucky and so were they. They were within yelling distance in the air and were on the ground within about a minute from canopy opening.

Planes flown: T-6G, T-28A, T-33, F-86D, F-86F.

Married Margaret Dec. 27, 1952. They have two children and two grandchildren.

Retired Aug. 1, 1975. Manager Physical Plant, Western Washington University. Resides in Bellingham, WA.

**JOE M. SNOW,** Maj. U.S.A.F. (Ret.), was born Aug. 15, 1917 in Mt. Home, PA. Entered the service Sept. 23, 1940. Served with the 340th B.G.

Flew B-25, B-17 and B-29.

Shot down over Enfioyville, North Africa near Tunis by 88th Panzer Division April 29, 1943. Bailed out over Mt. Erna Sept. 12, 1943. Night fighter got them. Lt. Blaney was killed in the mission.

Discharged April 1946. Recalled for Korean War and discharged again in November 1953.

Awards: Distinguished Flying Cross, Purple Heart, Air Medal with six OLC, N.A.T.O. with four Battle Stars and Presidential Citation with OLC.

Married to Adele Garland Blaney. They have two sons and six grandchildren.

Retired C.E.O., G&S Corporation until 1981.

Resides in Sarasota, FL.

**DONALD A. SOLBERG,** Lt. Col. U.S.A.F. (Ret.), was born Sept. 17, 1924 in Fairdale,

ND. Joined the Air Force Dec. 12, 1942. Graduated class 44-B, Douglas, AZ as a pilot. Transitioned in B-25s at Mathen, Sacramento, CA.

Assigned to 4th Combat Cargo E Squadron, 1st C.C. Group for operational training in C-47s, Louisville, KY and ordered to CBI theater via Azones, North Africa and Iran in August 1944. Stationed at Sylhet, Imphal, Dahazani and Hathazani, India doing air supply support for British army in Burma, Guam, Myithyena to Rangom and Chenghuny, China.

Bailed out in interior of Central China on Dec. 28, 1944 at 10 p.m. for lack of gas. Had flown a load of gas to a forward fighter strip which could not be spared for return.

Awarded Air Medal with three OLC and Distinguished Flying Cross with OLC.

Returned stateside 1945. Discharged May 1946. Active in Air Force Reserves until retirement. Achieved rank of lieutenant colonel.

Married to Marjorie C. Kepplin in 1948 and they have three children and seven grandchildren.

Earned B.S. in business administration in 1949. Employed by Unoral Corporation for 35 years and retired in 1987 as manager, benefits planning.

Member of the Hump Pilots Association and awarded honorary pilot wings in the Chinese Air Force. Currently resides in Yorba Linda, CA.

**NORMAN M. STEINHAUER (STEINY),** was born July 29, 1922 in Milwaukee, WI. Enlisted 1942 in U.S. Army Air Force. Took basic at Hondo, TX; flight engineering at Keesler Field; consolidated factory at San Diego; small arms at Camp Kearns, UT; gunnery at Laredo, TX.

Assigned to crew 898 in Salt Lake Ci and took crew training at March Field. Picke up B-24-J "Cherokee Belle" at Hamilto Field; P.O.E. at West Palm Beach. In a je accident and was left behind in Puerto Ric Caught up with the crew in Brazil and fle out of Africa and Italy.

Assigned to 15th A.F., 455th B.G., 741 B.S. Last location was near Cerignola, Ital "Dear crew member William (Wild Bi Delaney, volunteered his services and (hi for a special sortie", target Ploesti! Steinhau added that he had "properly thanked hi for that favor.

Shot down on 33rd mission, July 1944, in plane "Omiakinbak", target w Blechhammer Refinery near Breslau. Ca tured in Czechoslovakia. Interrogated Dulag Luft, Wetzlar and interned at Stala Luft IV - Gross Tychow. Room leader roo 13, barracks number three, Lager-A. A proaching Russians and injuries force evacuation by boxcar Jan. 30, 1945 to Stala Luft I, north compound III, block 303, roo four.

Liberated by Russians. Airlifted t Camp Ramp, "Lucky Strike". Returned o U.S.S. *Sea Tiger* to Newport News.

Discharged late 1945 at Truax Field WI.

Married over 50 years to Dorothy (ne Kaestnen). Children: Fred J. and Jeanne E

Retired plant manager of Alli Chalmers. Now resides in Claremont, NC

Original crew: R.M. Chadwick, pilc (deceased); H.O. Vralsted, co-pilot (de ceased); R.M. Crowley, navigator; J.M Grimm, bombardier; Writer, engineer; P. Daly, radio operator (deceased); W.A Delaney; H.A. Anderson aerial enginee (deceased); S.D. Schindler, aerial radio op erator (deceased); and A. Earl.

Awarded: three Air Medals, Purpl Heart, Presidential Unit Citation, ETO wit four Battle Stars; American Campaign, PO Medal; WWII Victory Medal; plus others.

**MILTON STEPHENS,** was born Dec. 1( 1929 in Columbus, GA. Entered service Sep tember 1948. Served with the 5th A.F., 8t F.W., 440th Sig., 48th T.F.W., 20th T.F.W 81st T.F.W.

Flew C-47, C-46 and C-118.

Served in Japan, Germany, England Korea, Vietnam and Saudi Arabia.

Was on C-46 returning in 1949 from Tachikawa, Japan to Itazuke, Japan with football team when the number one engine caught fire. Entire team bailed out. All safel jumped with only minor injuries resulting

Awarded: Bronze Star, Korean Cam paign, United Nations and various other with a combined total of 15.

Married Eileen and they have two sons John and Ricky and a daughter, Katrina, and two grandchildren. Resides in Ozark, AL.

Employed U.S. Army Aviation Center, Ft. Rucker, AL.

Discharged June 1974 with rank of SM/ Sgt.

## JOHN C. STEVENS,
was born April 26, 1924 in Rancocas, NJ. Entered the service April 14, 1943. Served with 443rd Troop Carrier Group, 2nd Troop Carrier Sqd.

Stationed at ShingBwyang, Burma, DimJam, and India.

Flew C-47 and C-46 as aerial engineer. Discharged Jan. 9, 1946 with rank of sergeant.

Married Dolores July 15, 1950. They have two children and two grandchildren.

Graduated from high school, trade school (auto research lab). Retired Dec. 30, 1980.

Enjoys walking, reading and occasional travel. Resides in Woodstown, NJ.

## CHARLES B. STRATTON (C.B.),
Col. U.S.A.F. (Ret.), was born May 25, 1929 in Nashville, TN. Joined the Air Force Jan. 16, 1951. Served with 508th Strat. Fighter Wing; 4080th Strat. Recon. Wing; 100th Strat. Recon. Wing.

On Jan. 2, 1962 while on a night celestial navigation mission in the U-2 at approximately 73,000 ft., the aircraft suddenly pitched up, stalled and began to break up. The ejection seat failed to fire through the canopy. Opened canopy manually and fell out. Pressure suit was inflated. Opened chute manually at approximately 50,000 ft. Chute caught in trees and Stratton spent the night above Pearl River Swamp near Picayune, MS. Rescued the following morning.

Awarded: Legion of Merit, Distin-

guished Flying Cross, Air Medal, Vietnam Service Medal.

Retired January 1982.

Married to Ann and they have two children. Resides near Spokane, WA. Spends most of his time hunting and fishing.

## RAYMOND H. STRAUTMAN,
was born Dec. 17, 1919. Entered service Aug. 18, 1941. Served with 450th B.G., 720th B.S. Tail gunner on "Miss Fortune" armorer.

Shot down April 25, 1944 in northern Italy by Me109 fighters. Plane was on fire. Bailed out, was wounded and sent to hospital. After about a month, sent to Stalag XVIII-C Markt Ponga, Austria then to Stalag VII-A. From there was sent to Stalag Luft IV (Gross Tychow) on Feb. 5, 1945 was started on a long forced march across northern Germany. Went from barn to barn, sometimes stayed outside. Crossed the Elb River, then put on boxcars and from there went to Hanover to Stalag IIB for a short time, then it was back to the Elb. Liberated by British on March 2, 1945.

Discharged Oct. 20, 1945 with rank of staff sergeant.

Awards: Distinguished Flying Cross with OLC, Purple Heart, Air Medal with seven OLC, and POW Medal.

Single and retired. Enjoys antique cars and parts.

## WRIGHT S. SWANAY,
2nd Lt., was born April 1, 1920 in Vonore, TN. Entered service July 3, 1942. Served with the 15th A.F., 459th B.G., 756th B.S.

Flew B-24 Liberator. Trained in Texas and Idaho.

On July 2, 1944, 39th mission, over Budapest, Hungary, was hit over target. Left formation and bailed out at 20,000 ft. Shot at by civilians while descending. Hid in a thicket from angry civilians. Evaded for three days before capture.

Sent to Stalag Luft III and two other prison camps. Forced on two death marches across Germany. Liberated April 29, 1945.

Awarded: Air Medal with two OLC, POW Medal, WWII Victory Medal, EAME Campaign Ribbon with two Battle Stars.

Married to Bernice and they have two sons, Bill and Russ and two grandchildren.

Wright was a computer specialist before retiring in 1972. He is active in local POW chapter and POW state department and authors both newsletters. He also does volunteer work at the local veterans' medical center.

Resides in Elizabethton, TN.

## DMITRI A. SWEETAK,
was born Oct. 24, 1916 in Nesquehoning, PA. Entered service Jan. 31, 1943 at Ft. Jackson, SC. Took cadet training at Santa Ana A.B., CA. Served

with 460th B.G., 760th B.S. in Africa and Italy.

Flew B-24, crew operating in the European Theater.

On March 1, 1944, on a mission to Austria, after dropping the bombs, they ran into heavy flak. Two engines were knocked out on the same side and cut the fuel lines, spilling gas all over the airplane. After trying to maintain altitude, and patching the fuel lines, they made it to Yugoslavia.

The order was given to bail out. The copilot and Sweetak bailed out at the same time but because of low altitude, Sweetak's chute did not open. He woke up hanging in a tree with his feet a foot and a half from the ground.

The airman was captured and taken to a field hospital for treatment. He managed to escape and that lead him to Tito's Partisans. Sweetak knew the language and became a member of their group.

Awards: Silver Star, Distinguished Flying Cross, Purple Heart, Presidential Group Citation with OLC, American Theatre Ribbon, EAME Ribbon with five Bronze Stars, and WWII Victory Medal.

Discharged Nov. 21, 1945 with rank of technical sergeant.

Married Helen Feb. 23, 1946. They have two children and three grandchildren.

Civilian education/employment: Tennessee Poly-Tech, Penn State. Retired in 1983. Resides in Pottstown, PA.

## WARREN W. SWENSON,
Maj. U.S.A.F. (Ret.), was born April 14, 1921 in Sister Bay, WI. Joined Army Air Force Jan. 30, 1943.

Piloted B-17s on combat missions in Europe. Also served in Italy, Japan, Okinawa, Korea and Guam. Served as U.S.A.F. command post pilot and electronics officer.

Bailed out of his burning C-123 35 seconds before explosion near Norfolk, VA on Oct. 10, 1958.

Awarded: Air Medal with OLC, 15th A.F., Korean Service Medal, Command Pilot, Guided Missile Insignia, and WWII Victory Medal.

Married to Genevieve Marquarot in Tomahawk, WI over 48 years ago. They have two children and one grandchild.

Retired U.S.A.F. major.

Member of Quiet Birdmen, 97th B.G. Reunion Association. Enjoys travel with Airstream travel trailer. Resides in Dade City, FL.

**E.A. THISTLETHWAITE,** was born Aug. 28, 1922 in Opelousas, LA. Joined U.S. Air Corps the summer of 1942. Made second lieutenant August 1943 at Eagle Pass, TX. Ordered to 339th F.G., 504th B.S.

First bailed out over the Mojave Desert in January 1944. Reached Fowlmere, England in March 1944.

Shot down May 30, 1944 over Magdeburg, Germany. Escaped about a month before the war ended.

Discharged in 1946 with rank of first lieutenant.

Thistlethwaite was a crop duster until 1988 when he retired and sold the business.

He raised three children (all are pilots).

The airman commented that his two jumps were very different. The Mojave one was from 250 ft. while the jump in Germany started from 27,000 ft. but he fell to about 1,000 ft. before pulling the cord. Says he is in good health and is "enjoying life." Also said that the fishing, hunting and golf are "excellent in Louisiana."

Resides in Washington, LA.

**ROBERT A. TITUS,** was born March 8, 1915 in Marion, IA. Drafted Aug. 19, 1941 into Field Artillery, but later became an Aviation Cadet. Completing navigator's training, October 1942, was assigned to the 95th BG (H) which, after training at Rapid City, SD, flew to England in April 1943.

On his third combat mission, May 29, crew had to bail out over Brittany. Helped by farmer Leandre Rochelle to escape to Swit-

zerland with two crew members (first Americans to arrive there).

While there worked in Intelligence with Allan Dulles (O.S.S.) Repatriated March 1944 and reassigned to the Ferry Command and later that year made two trips to England delivering A-26s.

En route to Australia by C-47 Jan. 18, 1945, engine failure resulted in ditching, 900 miles from the California coast, but all crew members were rescued from the raft.

After the war, resumed teaching career. Retired in 1981 after thirty-some years on faculty at Ohio State University.

**JOHN TOPOLSKI,** T/Sgt., was born June 24, 1919 in Chicago, IL. Inducted in 1942 and sent to Shepherd Field, TX. From there it was on to Steven's Hotel in Chicago for training in radio mechanics and code. From there he was sent to Harlingen, TX for a gunnery course and then to March Field in Riverside, CA where he was assigned to a B-24 crew.

Topolski went on to Shemya in the Aleutians where he took part in the campaign to take Kiska (held by Japanese). Pratt, KS was his next stop. There he trained on B-29's, then sent on to Boca Raton, FL for a course in the PP1 radar scope.

Served with the 11th A.F., 11th Command, 28th B.G., 4th A.F., 4th Command, 30th B.G., 21st Sqd., and the 20th A.F., 58th Wing, 40th B.G., 44th B.S. China, Burma, South Pacific and in India. In India radar was really pioneered for use in preparation for a bombing raid on Japan, from their home base in Chengtu, China.

Shot down over the Indian Ocean while returning from a mission in Singapore. Rescued by a British submarine after spending 29 hours in a Mae West. Remaining nine crewmen joined their squadron in the South Pacific and were assigned to Iwo Jima to fly escort missions for P-51 fighter planes.

After completion of this tour, Topolski was sent to Tinian where he was stationed when the A-bomb was dropped on Japan. The war was over. He was discharged from Camp McCoy, WI on Oct. 12, 1945 with rank of technical sergeant.

Awarded: Distinguished Flying Cross, Air Medal with four OLC, Good Conduct Medal, Bronze Star and a Presidential Unit Citation.

Married to Rita. They have six children and reside in Milwaukee, WI. Retired.

**DONALD R. TUOHY,** Cpl., was born Feb. 15, 1926 in San Francisco, CA. Enlisted Feb. 9, 1944. Served in the 2nd A.F. and the 4th A.F., where he completed training as an aerial gunner on the B-24 bomber. Then stationed at Ft. Worth, TX where he took training on the B-32 very heavy bomber. Closed his military career in the 21st B.S., 501st B.G.,

a B-29 outfit stationed in the Marianas Islands.

Involved in a mid-air collision of two B-24 bombers while on a practice mission in South Dakota. The plane had just taken off when they joined the circling formation of B-24 bombers over Cortez, CO when another bomber did the same. They lost both top halves of their horizontal stabilizers and rudders and the plane went down.

The other B-24 managed to make it back to the base at Pueblo, CO. The navigator broke his leg, but there were no other injuries.

Awarded: Asiatic Pacific Campaign Medal, Good Conduct Medal, American Theatre, and WWII Victory Medal.

Discharged April 26, 1946 with rank of corporal.

Earned a B.A. degree in anthropology, sociology and geography at San Francisco State University in 1952. Went on to graduate school at the University of Washington and the University of Arizona; and was awarded an M.A. degree in anthropology from the University of Nevada in Las Vegas in 1978.

Since the late 1950s, Tuohy has pursued a professional career in prehistoric archaeology, while serving most of his time as curator of anthropology at the Nevada State Museum in Carson City, NV.

Married Lila Williams in 1964 and they had one set of fraternal twins (one boy and one girl), and another daughter in 1960s. They have one granddaughter so far.

**ROBERT C. TWYFORD,** S/Sgt., was born in Butler, PA. Entered the service Nov. 19, 1940. Served with the 901st H.A.M. Co., 467th B.G., 791st B.S. in Rackheath, England.

Flew B-24s.

Aircraft took a direct hit over Berlin, Germany March 18, 1945 on 24th mission.

Bailed out of B-24 near Landsburg, Germany. Russians picked them up and took them to Poznzn, Poland, then to Lublin to Poltova, Russia and back to England.

Awards: Air Medal with OLC, ETO Medal, WWII Victory Medal, American Theatre Campaign and Good Conduct Medals.

Married to Dorothy and they have two children, Bob and Linda (both deceased). Retired accountant. Enjoys golf, fishing, hiking and travel.

**RICHARD G. VOSS, M.D.,** was born March 7, 1929 in Osborne, KS. Graduated from Colorado College in 1956 and from Washington University Medical School in St. Louis, MO in 1960. Entered U.S.A.F. in November 1950. Served with the 49th B.G., 8th B.S. (Black Sheep), Training Command, Luke Air Force Base, James Connelly AFB, Williams AFB, K-2 Korea. Was pilot and jet pilot instructor (F-84), gunnery.

Shot down by ground fire July 14, 1952 on front line at Ma-Bong-NI, North Korea, on a close support mission. Hit on bomb run. Taken POW after parachuting out of F-84E. Suffered a broken leg and third degree burns. Released September 1953. Had been declared killed in action. Repatriated September 1953.

Discharged Dec. 14, 1954 with rank of first lieutenant.

Awarded: Korean Service Medal, United Nations Service Medal, National Defense Service Medal, Air Medal, Distinguished Flying Cross and Purple Heart.

Married over 36 years and has two children. Voss became a medical doctor. Resides in Ft. Collins, CO.

**HOWARD B. WALSH (BEN),** Col., was born Sept. 25, 1919 in Hampton, IA. Entered service Nov. 28, 1940. Served with 389th B.G., 14th Bomb Wing, 458th B.G. in Tunisia, Benghazi, Libya, Shipdham, Hethel, Horsham St. Faith, England.

Planes flown: B-17, B-24, P-47, P-51, A-20, B-25, B-47, T-33, C-130, etc. during 30 years of flying.

Walsh was a career military officer. Discharged Jan. 11, 1970 with rank of colonel.

Married Barbara June 1, 1945. They have two children and three grandchildren and reside in Santa Barbara, CA.

Director of National Alliance of Business.

**RICHARD L. WANN,** was born in Elwood, IN. Grew up on a family farm near Summitville. Enlisted in Army Air Force Oct. 15, 1942. Graduated as a pilot, 2nd lieutenant, in May 1944.

Served with the 446th B.G., B-24, "Lil' Snooks", as crew commander. Shot down Feb. 3, 1945 on mission to Magdeburg, Germany.

Bailed out near Zweibrucken. Landed in no-man's land between front lines and evaded capture for seven days.

Placed in Heppenheim Prison Hospital due to injuries. Wann commented that the German commandant expressed that he hated Americans' guts and treated them accordingly with physical abuse, threats, lack of food, care and medical attention.

Wann was liberated March 27th by the 7th Army. Awarded Purple Heart.

Married to Ruth Procter of Elwood, IN. Graduated from Purdue University with a B.S. and M.S. Received honor of "Purdue Old Master" in 1965.

Civilian career spent in education and engineering management. Retired as division manager, (30 years), for Firestone, Akron, OH in 1982. Returned to Elwood, IN in 1987.

**BOB WEILER,** was born Sept. 26, 1919 in New York City, NY. Joined Air Force September 1943. Prior field artillery and tank destroyer force. Served with the 61st B.S., 39th B.G., 314th Wing on Guam. Locations: Tyndall AFB, Great Bend AFB, KS, and Salina, KS.

Qualified to be a Caterpillar on June 1, 1945, as left blister gunner on a B-29 in the 314th Wing based on Guam. They were on a daylight mission against Osaka. Before the bomb run they lost air pressure in the number three engine and feathered. They fell behind the squadron but put their bombs on the target, picking up a lot of flak.

To evade fighters, they flew into the smoke clouds of burning factories, etc. and came out of the smoke upside down and

higher due to the updrafts. It was "a really wild ride!"

As they were crossing the coast coming out they couldn't keep number three in feather and it threw the prop, cutting through the lower two-thirds of the fuselage. With two engines out and only the left Aileron responding, the aircraft kept them in the air until they were near Sofu Gan (Lot's Wife), a rock halfway to Iwo Jima.

They bailed out at 900 ft. and were picked up the next day by an American submarine, the Tinosa. The crew was transferred two days later to the Scabbard Fish in the mouth of Nagasaki Harbor at night for the ride back to Guam. The flight engineer died on the mission. His chute failed to open and was not found.

In addition to bailout, and participation in later Tokyo fire raids, Weiler spoke to General Tooey Spaatz, while on guard duty at the aircraft.

Discharged Dec. 2, 1945 with rank of sergeant.

Earned B.A. degree in economics at Hofstra University, Hempstead, NY. Was chief control agent for airlines and later owned a travel agency. Retired in 1977 but is still active in scouting, aviation art (does aircraft industrial painting part time), the Air Force Association and the Military Order of Purple Heart.

Married for over 45 years and has three children, four grandchildren and one great-grandchild. Resides in Sarasota, FL.

**IRA P. WEINSTEIN,** 1st Lt. U.S.A.C., was born June 10, 1919 in Chicago, IL. Enlisted Aug. 7, 1942. Was Aviation Cadet at Ellington Field, Childress, TX. Specialty: bombardier - D.R. navigator, Pathfinder crew flying B-24 Liberators. Assigned to 445th B.G., 702nd B.S., Tebinham, England.

Flew 25 missions. Shot down on Sept. 27, 1945 on the famed Kassel, Germany raid. Group lost 28 of 36 planes.

Weinstein bailed out through the nose wheel hatch. His parachute harness caught on bomb sight. Chinned himself back into the airplane, unhooked his chute and finally got free of the plane at about 2,000 ft. He was captured and interrogated at Oberusal. Interred at Stalag Luft I, Barth, Germany.

Discharged December 1945 with rank of first lieutenant.

Awarded: Distinguished Flying Cross, Air Medal with OLC, Purple Heart with OLC, Presidential Citation, Distinguished Unit Citation, POW Medal, Good Conduct Medal and American Theatre Victory Medal.

President and chief executive officer of Schram Advertising Agency.

Married to Norma Weinstein. They have two children and two grandchildren. Resides in Glencoe, IL and Palm Beach, FL.

Member of: Ex-POW/MIA Assoc., Caterpillar Club, 8th A.F. Historical Soc., Second Air Division Assoc. 8th A.F., Jewish War Veterans, 8th A.F. Memorial Museum Foundation, B-24 Liberator Assoc., and the Air Force Assoc.

## WILLIAM DUNCAN WELDER, JR.,

Capt., was born Oct. 19, 1930 in Houston, TX. Joined U.S.A.F. Sept. 1, 1951. Served with the 36th F.B.S., 5th A.F. Hq. Advance in Korea and San Antonio.

While on a routine training mission in an F-84 out of Luke AFB in Arizona, July 1, 1953, Welder was forced to eject due to failure in the aileron boost hydraulic system. The ejection occurred at approximately 30,000 ft. at air speed in excess of 500 mph. The airman fractured his right arm and sustained many cuts, bruises and cactus spines.

He landed approximately three miles north of the town of Aguila near the old Butterfield stage route and was picked up by a couple of local ranchers who happened to be in the area. They took him to the airstrip north of Aguila and there he was met by the flight surgeon, an ambulance and the search team.

Because of the high altitude and the long free-fall no one in the flight saw his parachute open and he was reported as lost with the plane. Welder was temporarily blinded in his right eye and did not pull the "D" ring until he saw the smoke from the plane. After pulling the chute, he was on the ground in about 30 seconds, landing only 20 yards from the plane impact crater.

Discharged from U.S.A.F. Reserves Oct. 3, 1967 with rank of captain.

Married Wanda Simmons and they have six children and seven grandchildren.

Retired June 30, 1989 from Rohm and Haas Chemical Company. Now a rancher in Liberty County, TX.

## LEW WICKENS,

Lt. U.S.A.F. (Ret.), was born Jan. 8, 1924 in Liberty, MO. Enlisted April 9, 1942. Service in: Ft. McArthur, Camp Haan, Santa Ana AFB, Gary, Minter and Williams Fields, Seymour Johnson, NC, Woodchurch, England, Bayeaux and Rheims, France and Frankfurt and Barth, Germany, Luke AFB.

Combat with: 410th F.S., 373rd F.G., 19th T.A.C., 303rd F.W., 9th A.F.

Memorable experiences: chasing V-1 rockets in southwest England, then helping run Jerry from the beachhead across from France to the Mouselle as a "flying infantry-man."

Biggest mistake: volunteering TWICE on Sept. 15, 1944. The result was that instead of spending the night on 100-hour combat R&R in Scotland, he took Luftwaffe billeting in an old barn outside Saarbrucken Germany. Said Wickens: "Considering the choice of a deck-level, flaming Thunderbolt hit by flak and an accelerated chute test, the silk proved quick n' cool."

Awarded: Purple Heart, Air Medal with four OLC, POW Medal, Presidential Unit Citation, ETO with three Battle Stars and American Theatre.

Graduated 1949 with B.S. at U.C.L.A. Retired after 35 development years in the frozen food industry.

Married to Lois. They have two sons and four grandchildren and reside in Craig, MO.

Best part lately: sharing in one of only two Stalag Luft I ex-roommates reunions, celebrating a 45th year Liberation Day anniversary. Fourteen of the original 18 reluctant chutists remain "The Hungry Hollow Bunch", rejoined and rejoiced.

## S.J. WIGLEY (JAY),

Maj. U.S.A.F. (Ret.), was born Oct. 31, 1920 in Maysville, OK. Enlisted U.S. Army Air Corps June 28, 1940. Instructor - Technical Training Command 1941-43. Flying school 1944-45 at Santa Ana, Tulare and Bakersfield, CA and Marfa, TX.

Troop Carrier 1945-47 in U.S., Japan and Korea. Light bombers, Tinker AFB 1948-50. Strategic Air Command 1951-63 except for 1953 in Korea as V.I.P. pilot.

Flew: C-46, C-47, C-54, C-82, C-119, B-26, and B-36.

Forced bailout occurred in April 1956 on a mission from Biggs AFB in Texas to Moses

Lake Washington to pick up a C-47 aircra[ft] They were to ferry to Moses Lake in a co[n]verted Douglas B-26 bomber. Lost radio co[n]trol in unforseen snow and freezing rain a[nd] got lost. Order came to bail out when th[ey] started running out of fuel.

Wigley descended into an area of free[z]ing rain and had to shake his parachu[te] canopy so sheets of ice would slide off. Al[so] broke ice from his face, hands and clothin[g]. Two minutes after bailout he heard the ai[r]craft hit and explode while he was still se[v]eral thousand feet up. He continued to fig[ht] the ice until he was below the ice level an[d] into rain. He landed in a meadow in tw[o] inches of water and lots of mud.

Honored as SAC Pilot of Month in Ap[ril] 1959 for flying a C-47 40 miles after losi[ng] second engine and landing safely with a loa[d] of 14 passengers.

Retired Jan. 1, 1964.

Awarded: several commendation letter[s] 10 service medals with OLC's and Batt[le] Stars.

Married to Virginia and they have thre[e] children and two grandchildren.

Retired building contractor, education[al] construction, estimating engineer.

Resides in Oklahoma City, OK.

## THOMAS C. WILCOX,

T/Sg[t.] U.S.A.A.C., was born May 27, 1923 in Young[s]town, OH. Drafted into military service Fe[b.] 16, 1943. Took radio operator mechanic trai[n]ing at Scott Field, IL; gunnery training [at] Buckingham AAF Ft. Myers, FL, was as[s]igned to the 344th B.G., 496th B.S. at Lak[e]land, FL, where he received his oversea[s] training in the Martin B-26 Marauder.

Flew with plane via the southern rou[te] to base at Stansted Airfield near Bisho[p] Stortford, England.

Shot down on 67th mission over Venlo, Holland Sept. 23, 1944. With help of Dutch Underground, escaped on Nov. 24, 1944.

Returned to U.S. and entered pilots training at SAAC, TX, completed ten week training course when war with Germany ended. Pilot training was discontinued. Wilcox was discharged July 15, 1945 with rank of technical sergeant.

Awarded: Purple Heart, Air Medal, two Silver Stars, two Bronze Stars with OLCs.

Earned pastor's certificate at Cornus Hill Bible College, Akron, OH. Retired January 1980. Currently teaches Bible School, is a part time realtor and enjoys boating, fishing and shelling.

Married E.L. (Pat) Shepard Aug. 12, 1945. Children: David T., Rebecca E., Deborah G. and Melody E. Grandchildren: Michele and Mark Wilcox, Seth and Justin Gang, Rayne, Brianna and Dani Smith. First wife died May 1985. Remarried May 1987, Mary Hudson. Resides in Ft. Myers Beach, FL.

**EUGENE A. WINK, JR.,** Lt. Col. U.S.A.F. (Ret.), was born Oct. 19, 1920 in Biloxi, MS. Pilot training in Texas, class 42-K. Graduate U.S. Military Academy, class January 1943. Commissioned second lieutenant U.S. Army, West Point, NY.

Sent to England with 365th F.G. after P-47 training at Richmond, VA. Bailed out over northern France March 2, 1944 after fighting FW-190's. Evaded enemy in France with help of Etienne Capron, a courageous Frenchman and World War I hero from Graincourt-les-Havrincourt. Crossed Pyrenees Mountains into Spain, arriving England April 17, 1944.

Performed operational and staff duties in U.S.A.F. Flew air rescue missions in Korean War and KC-135 sorties with SAC during Vietnam War.

Became a banker after USAF retirement September 1969. Retired senior vice president, Frost Bank Corporation 1985.

Married 1947 to Dr. Irma June Wink. Children: Sue Karen Wink, M.D., Robin Wink, and Bruce Wink. Resides in San Antonio, TX.

**BENNETT WINTERS,** Maj. U.S.A.F. (Ret.), was born April 3, 1939 in Port Townsend, WA. Entered officer training school Feb. 16, 1966.

Service in: Vietnam 8th T.F.W. (Wolfpack), Germany, 52nd T.F.W., 12th A.F. Hq., Austin, TX.

Most memorable mission: On Nov. 22, 1980, while performing duties as a flight examiner for 12th A.F. Hq., Winters was administering a flight evaluation to the chief of Stan/Eval, Reno, NV National Guard flying F-4 Phantoms when the aircraft burst into flames. Unable to extinguish the engine fire and realizing the aircraft could explode at any moment, the plane was turned towards the mountains and away from the city of Reno. Both crew members successfully ejected. The alert examinee quickly turned on his side looking cameras and photographed the entire ejection sequence. Rare or perhaps the only time an actual USAF ejection was photographed.

Awarded: Distinguished Flying Cross, Air Medal with 13 OLC, Air Force Outstanding Unit with OLC, Combat Readiness Medal with three OLC, Rep of Vietnam Gallantry Cross with Palm.

Retired Oct. 2, 1985 with rank of major. M.A. degree Central Michigan University, B.S. degree Washington State University.

Married to Susan Nov. 27, 1982 and they have three children. Resides in Portland, OR.

**WILLIAM H. WISE,** Maj. Gen. U.S.M.A. (Ret.), was born May 6, 1909 in Pontiac, IL. Enlisted 1927, U.S.M.A. 1930-34, commissioned 1934. Served with 3rd Attack Grp., 8th F.G., Hq. A.A.F., 32nd T.A.C., JTF 3, WADF, NEAC, 37th Air Def. Div., 32nd Air Def. Div., Hq. NORAD.

Flew mostly trainers, starting with PT-3, fighters from P-1 to F-104, misc. bombers and cargo.

Jan. 12, 1938, PB-2A, two-place fighter,

Airman Riley Scruggs in back seat. At 1,000 ft. in traffic pattern, Langley Field, VA. Engine threw con rod and caught fire.

Wise's first reaction was to dead-stick into field. Cut switches and looked down at spot where Abie Waller had been killed a month earlier doing the same thing. Pulled nose up, signaled Scruggs to jump. Got out, pulled ripcord, hit ground. Airplane trailing flames, dug 12 ft. hole in ground nearby. Was first lieutenant at the time of jump.

Discharged March 31, 1966 with rank of major general.

Awarded: Distinguished Service Medal, Legion of Merit with OLC, Bronze Star Medal, Air Medal with OLC, CR with OLC, all theatres, etc.

Married to Anna Powell Burns. Two children and three grandchildren. Retired in Colorado.

Enjoys trail riding, skiing, and trout fishing. Plans to move to A.F. Village.

**MAX J. WOOLLEY,** Capt. U.S.A.F., was born Dec. 19, 1916 in Ryegate, MT.

Entered service May 8, 1942. Attended flight schools at Santa Ana, Oxnard, Lemoore, CA, Williams Field, AZ. Commissioned and rated pilot July 28, 1943. Assigned to 364th F.G., 384th F.S. Tactical training at Muroc (Edwards), Van Nuys, Ontario, CA.

His P-38 fighter was hit by anti-aircraft fire while strafing a German airfield near Charleroi, Belgian on 45th mission June 27, 1944. Woolley was shot by enemy ground troops descending in parachute.

A Belgium family hid and fed him and attended to his wounds while MIA. Collaborator informed secret police. All were placed on Gestapo list as "Third Reich" enemies. All were saved from incarceration, or "whatever", by Allied Forces two hours before order effective.

Awarded: Distinguished Flying Cross, Air Medal with three OLC, Purple Heart, and Presidential Group Citation.

Discharged Aug. 19, 1946 with rank of captain.

Married Blanche Brown Dec. 5, 1944. They have one son, Max.

Earned B.A. from Chapman College and B.P.A. from Chouinard Art Institute in Los

Angeles, CA. Commercial artist with paintings hanging in numerous art galleries, private collections and Kennedy Center.

**MARSHALL D. WORD,** was born June 28, 1915 in Elk City, OK. Graduated from Oklahoma University Law School in 1940. Entered service as Aviation Cadet Aug. 28, 1941.

Received pilot wings and commissioned second lieutenant April 18, 1942, Stockton Field, CA. Instructed BT-13s Gardner Field, CA, until Aug. 7, 1943; B-24 instructor one year at Kirtland Field, NM; then sent to Italy Aug. 1, 1944 with 451st BG, 727th Sqdn., operations officer.

Shot down Osweicim mission Dec. 26, 1944; M.I.A. 23 days. Yugoslav Partisans aided return to Italy; completed 26 missions earning Distinguished Flying Cross and Air Medal with Clusters; was squadron commander of the 725th Sqdn. prior to return to U.S.

Remained in Air Force Reserve serving in SAC and later in OSI for summer tours; retired 1976 as lieutenant colonel.

Except for military service, practiced law since 1940, retiring 1980. Married Elta Hibler, May 16, 1942; two sons, Bill and Richard, five grandchildren. Resides in Arnett, OK.

**MARCUS H. WORDE,** Lt. Col. U.S.A.F. (Ret.), was born September 5, 1920 in Morrison, TN. Entered aviation cadet training September 1941 in Atlanta, GA. Commissioned and rated a pilot at Victoria AAF, TX, April 29, 1942.

Assignments included instructor of B-24 instructor pilots, 2nd A.F.; B-29 aircraft commander and lead crew, 20th A.F.; Army and Air Force recruiting deputy for Mississippi, 3rd Army; logistics staff officer, Hq. USAF-Pentagon; commander 13th B.S., Far

Eastern Air Forces; assistant wing deputy commander for maintenance, Strategic Air Command.

Air combat: Shot down on 16th B-29 mission over Yokohama, Japan, May 29, 1945. Bailed out and held at Kempei Tai headquarters and Omori Prisoner of War Camps until Aug. 29, 1945.

Awarded: Purple Heart, POW Medal, Air Medal with three OLC, Air Force Commendation Medal with three OLC and Presidential Unit Citation with OLC.

Member The Retired Officers Association, Ex-POWs, and Caterpillar Association. Professional member American Engineering Association.

Married to Faye Smotherman of Erick, OK. They have three children and four grandchildren.

**JOHN M. WYLDER,** was born Dec. 17, 1920 in Kansas City, MO. Enlisted in the U.S.A.A.F. Aug. 27, 1942 in Los Angeles, CA.

Bombardier training: Carlsbad, NM, 43-18, December 1943. Commissioned second lieutenant June 1, 1944, 13th A.F., 5th BG, 23rd Bomb Sqdn. Flew mission when 5th BG, 23rd Sqdn. sunk Japanese cruiser "Kuma Natori" class Battle of the Philippine Sea.

On 18th mission to a "milk-run" target shot down Nov. 1, 1944. M.I.A. Floated off

shore Bacolad Air Field. Strafed in shark [in]fested waters. Rescued by Philippine guer[ril]las in broad daylight. Fought with gueril[la] under Lt. Col. Cirilo B. Garcia. Walked 1[00] miles to submarine rendezvous area. R[es]cued by USS *Hake* (SS-256) Dec. 5-16, 19[44]. Depth charged, dive bombed at scope dep[th] and chased on surface at night while runni[ng] a mine field at flank speed.

Awarded: Purple Heart, President[ial] Unit Citation, Asiatic-Pacific Service Med[al] with four Clusters, Air Medal with Oak L[eaf] Cluster, Philippines Liberation Medal a[nd] Star.

Wife, Kay; children: Linda and Robe[rt]. Noise control consultant and distributor [of] noise control materials, Van Nuys, CA. R[e]sides in Van Nuys, CA.

**BRUCE A. YUNGCLAS,** was born May [?] 1924 in Webster City, IA. Entered serv[ice] Dec. 20, 1943 in Des Moines, IA. Served [in] Army Air Corps as radar operator on Tini[an] Island.

Bailed out over Japan after being hit b[y] flak over Yokohama May 29, 1945.

Awarded Air Medal with two OL[C]. Achieved rank of technical sergeant.

Married to Patty Olsen June 20, 194[?]. They have three children and seven gran[d]children. Resides in Webster City, IA. Retire[d] farmer.

*Art Zander and Hugh Harries, both crew members on a B-24, are reunited after 45 years. They parachuted from their burning plane, and were later captured and held as POWs in Germany.*

Printed in the USA
CPSIA information can be obtained
at www.ICGtesting.com
LVHW081110210924
791744LV00005B/64

9 781563 110313